空气动力学基础与飞行原理

Aerodynamic Fundamentals and Flight Theory

主　编　王　磊　顾　菘　陈　璐

副主编　韩　雷　孙瑜珮　刘旭辉　夏雪梅

参　编　孟柯生　张　驰　麻肖妃　赵　蓉

北京理工大学出版社

BEIJING INSTITUTE OF TECHNOLOGY PRESS

内容提要

本书基于中国民用航空器维修人员考试中关于空气动力学和飞行原理学习的具体要求，系统地介绍了空气动力学的基本原理和航空飞行理论。本书注重厘清基本概念和原理并进行定性分析，力求做到概念清楚、理论正确、知识点全面、突出实用性、注重理论和实践相结合。本书分为8个模块，包括大气环境、大气基本原理、机翼形状和参数、空气动力、高速飞行、飞行原理基础、飞行稳定性、飞行操纵性。

本书可作为相关院校航空器维修专业的教材，也可作为航空器在职维修人员和其他技术人员的参考用书。

图书在版编目（CIP）数据

空气动力学基础与飞行原理：英汉 / 王磊，顾菘，
陈璐主编. -- 北京：北京理工大学出版社，2023.8
　　ISBN 978-7-5763-2724-3

　　Ⅰ.①空…　Ⅱ.①王…②顾…③陈…　Ⅲ.①空气动
力学－英、汉②飞行原理－英、汉　Ⅳ.①V21

中国国家版本馆CIP数据核字（2023）第150357号

责任编辑：王梦春	文案编辑：辛丽莉
责任校对：周瑞红	责任印制：王美丽

出版发行 / 北京理工大学出版社有限责任公司	
社　　址 / 北京市丰台区四合庄路6号	
邮　　编 / 100070	
电　　话 / （010）68914026（教材售后服务热线）	
（010）68944437（课件资源服务热线）	
网　　址 / http：//www.bitpress.com.cn	
版 印 次 / 2023年8月第1版第1次印刷	
印　　刷 / 河北鑫彩博图印刷有限公司	
开　　本 / 787 mm×1092 mm　1/16	
印　　张 / 18.5	
字　　数 / 383千字	
定　　价 / 82.00元	

前　言

党的二十大报告指出："实施科教兴国战略，强化现代化建设人才支撑。"

高等教育不仅是培养高素质技术人才的重要途径，更是推动社会经济发展的关键力量。为适应新时代的发展要求，满足航空类专业人才培养需要，本书力求为广大学生提供全面、系统、实用的知识与技能。

本书编写力求全面覆盖航空相关专业领域理论和实践知识，探讨和分析航空领域各专业的前沿技术和趋势，结合实际案例，让广大学生真正掌握行业核心技能。

本书依据 2020 年中国民用航空局规章《民用航空器维修人员执照管理规则》（CCAR-66R3），并按照 2020 年中国民用航空局咨询通告《航空器维修基础知识和实作培训规范》（AC-66-FS-002 R1）的要求编写，基于航空器维修人员考试中关于空气动力学、飞行原理学以及航空维修技术英语等级测试的具体要求选取内容，采用双语的形式系统地介绍了空气动力学的基本原理和航空飞行理论。

本书结合读者实际学习需求和学习特点，注重对接行业标准，参照中国民用航空局航空器维修人员执照考试的知识内容和能力要求，突出实用性。本书按照任务要求和知识点组织，打破传统教学大纲模式，转向具体实践和操作过程；重点和难点按照任务或知识点的方式组织，循序渐进，力求做到对重点和难点的准确阐述和练习反馈，同时，重点和难点在不同的教材内容之间横向、纵向关联，通过不同任务和知识点达到细化巩固的作用。

本书采用"模块—任务"的编写形式。每个任务包含导读、课文正文、单词表、课后练习、拓展阅读等部分，注重教材的实用性。导读部分通过介绍任务内容对课程进行导入；课文正文采用中英文对照的方式进行介绍；单词表给出文中的专业英语词汇的释义，全书专业词汇量超过 800 词；课后练习根据学习目标对内容进行编排，可用于预习思考、课堂提问和复习总结；拓展阅读针对课文内容进行扩充和补充说明，可作为学有余力的学生的学习资料及课堂内容的辅助参考学习资料。

本书由成都航空职业技术学院王磊、顾菘、韩雷、孙瑜珮、刘旭辉、麻肖妃、赵蓉，安徽交通职业技术学院孟柯生，北京飞机维修工程有限公司（Ameco）陈璐，四川航空股份有限公司张驰，以及航空工业成都飞机工业（集团）有限责任公司夏雪梅共同编写。本书在编写过程中，得到了航空相关单位以及航空维修领域专家的大力支持，并参考了国内外同行的相关文献，在此一并表示感谢。

由于编写时间仓促，编者水平有限，书中难免存在疏漏之处，恳请各位读者批评指正。

编　者

目 录

Contents

06 Module 6　Flight Fundamentals

07 Module 7　Flight Stability

08 Module 8　Flight Maneuverability

模块1
01
Module 1

大气环境
Atmosphere

Contents

1) Composition of the atmosphere

2) Layers of the atmosphere

3) Properties of the atmosphere

学习内容

1）大气的组成

2）大气层的结构层次

3）大气的物理性质

任务 1　大气的组成
Task 1　Composition of the Atmosphere

Contents

Composition of the atmosphere

Learning Outcomes

1) Understand the composition of the atmosphere

2) Solve the aerodynamics problems by the composition of the atmosphere

3) Cultivate professional qualities of rigor, carefulness, and ability to express, coordinate, and communicate effectively

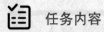

任务内容

大气的组成

任务目标

1）理解大气的组成
2）运用大气的组成解决相关的空气动力学问题
3）培养严谨、细心的职业素养，以及有效表达、协调和沟通的能力

Learning Guide

Although the atmosphere is transparent, it is composed of many substances. What chemical components and substances do you think the atmosphere will contain? What other substances are there? Which of these substances that make up the atmosphere will have an impact on human production, life, and aviation activities?

课文

Composition of the Atmosphere
大气的组成

The atmosphere is a mixture of many gases(Fig. 1-1). The atmosphere is mainly composed of nitrogen(N_2) and oxygen(O_2). By volume, nitrogen accounts for about 78% and oxygen 21%. The remaining 1% is argon, carbon dioxide(CO_2), neon, helium, krypton, hydrogen and other gases. Besides gases, the atmosphere also contains water vapor and dust particles.

Composition of the Atmosphere

大气是多种成分组成的混合气体（如图 1-1 所示）。大气主要由氮气和氧气组成。按体积计算，氮气约占 78%，氧气约占 21%。剩下的 1% 是氩、二氧化碳、氖、氦、氪、氢和其他气体。除气体外，大气中还含有水蒸气和尘埃颗粒。

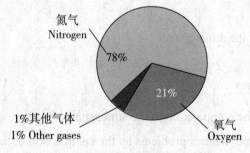

Fig. 1-1 Composition of the atmosphere
图 1-1 大气的组成

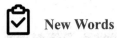 **New Words**

atmosphere	['ætməsfɪə(r)]	n.	大气，大气层，大气圈
nitrogen	['naɪtrədʒən]	n.	氮，氮气
oxygen	['ɒksɪdʒən]	n.	氧，氧气
carbon dioxide	['kɑr‚bən daɪ'ɒksaɪd]		二氧化碳
vapor	['veɪpə(r)]	n.	蒸汽，水汽
dust	[dʌst]	n.	沙土，尘土，灰尘，尘埃，粉尘，粉末
particle	['pɒtɪkl]	n.	颗粒，微粒，粒子
volume	['vɒljuːm]	n.	体积，容积，容量

 Q&A

The following questions are for you to answer and assess the learning outcomes.

(1) Understand the composition of the atmosphere.

(2) What are the two main gases in the atmosphere?

(3) What components in the atmosphere may cause bad influence to the flight activity?

(4) How to protect the relevant aspects from the bad influence in atmosphere?

 Extended Reading

What Is the Atmosphere?

The atmosphere is a mixture of gases that surrounds the planet. On the Earth, the atmosphere helps make life possible. Besides providing us with something to breathe, it shields us from most of the harmful ultraviolet (UV) radiation coming from the Sun, warms the surface of our planet by about 33 ℃ via the greenhouse effect, and largely prevents extreme differences between daytime and nighttime temperatures. The other planets in our solar system also have an atmosphere, but none of them have the same ratio of gases and layered structure as the Earth's atmosphere.

Nitrogen and oxygen are by far the most common gases; dry air is composed of about 78% nitrogen and about 21% oxygen. Argon, carbon dioxide, and many other gases are also present in much lower amounts; each makes up less than 1% of the atmosphere's mixture of gases. The atmosphere also includes water vapor. The amount of water vapor present varies a lot, but on average is around 1%. There are also many small particles, solids and liquids, "floating" in the atmosphere. These particles, which scientists call "aerosols", include dust, spores and pollen, salt from sea spray, volcanic ash, smoke, and more.

任务 2 大气层的组成
Task 2 Layers of the Atmosphere

Contents

1）Layers of the atmosphere
2）Troposphere
3）Stratosphere

Learning Outcomes

1）Understand the composition of the atmosphere layers
2）Understand the physical properties of the atmosphere at different altitudes
3）Solve the aerodynamics problems by the layers of the atmosphere
4）Cultivate professional qualities of rigor, carefulness, and ability to express, coordinate, and communicate effectively

任务内容

1）大气的结构层次
2）对流层
3）平流层

任务目标

1）理解大气层的组成
2）理解不同高度大气层的物理性质
3）运用大气层的组成解决相关的空气动力学问题
4）培养严谨、细心的职业素养，以及有效表达、协调和沟通的能力

Learning Guide

The weight of the entire atmosphere is approximately 5,000 trillion tons. The density of the upper layer is much smaller than that of the lower layer, and the higher the density, the thinner it becomes. If the density of air at sea level is taken as 1, the atmospheric density at 240 km altitude is only one in ten million of it. At an altitude of 1,600 km, it becomes even rarer, only one billionth of its size. 90% of the mass of the entire atmosphere is concentrated in space within 16 km above sea level.

 课文

Layers of the Atmosphere
大气的结构层次

The thickness of the atmosphere is over 1,000 km. According to the variation of atmospheric temperature with altitude, the atmosphere from bottom to top, is divided into troposphere, stratosphere, mesosphere, thermosphere and exosphere(Fig. 1-2).

大气层的厚度超过 1 000 km。根据大气温度随高度分布的特征，可以将大气自下而上分为对流层、平流层、中间层、热层和外逸层（如图 1-2 所示）。

Layers of the Atmosphere (1)

Layers of the Atmosphere (2)

Layers of the Atmosphere (3)

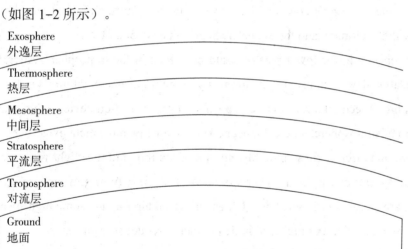

Fig. 1-2　The layers of the atmosphere
图 1-2　大气的结构层次

1. Troposphere
1. 对流层

Troposphere is the layer closest to the earth surface. In the middle latitude of the earth, its top layer is about 11 km away from the ground, and it is about 17 km high at the equator of the earth, and is lower at the poles, about 7-8 km.

Due to the gravity, 3/4 of the total mass of the atmosphere and all water vapor are concentrated in this layer, the troposphere is the most complex layer of weather change, including clouds, rains, hails and other phenomena.

At this altitude, there are horizontal and vertical airflow, forming horizontal and vertical gusts; The physical parameters (pressure, density, temperature and speed of sound) of the atmosphere decrease with the increase of altitude. This layer is the main area of aircraft activities.

对流层是地球大气层中最靠近地面的一层。在地球的中纬度，其顶层距离地面约为 11 km，在地球赤道附近约为 17 km，两极较低，为 7~8 km。

由于重力作用，大气总质量的 3/4 和所有水蒸气都集中在这一层，因此对流层是天气变化最复杂的一层，包括云、雨、冰雹和其他现象。

在这个高度，有水平和垂直气流，形成水平和垂直阵风；大气的物理参数（压力、密度、温度和声速）随着海拔的升高而降低。该层是飞机活动的主要区域。

2. Stratosphere

2. 平流层

The stratosphere is located on the troposphere, with the top layer about 50 km from the ground.

In the lower half of the stratosphere (about 20 km below), the temperature of the stratosphere is almost unchanged with the altitude, and the annual average value is −56.5 ℃.

Aircraft generally operate in the lower part of stratosphere and in the troposphere. Aircraft without cabin pressurization device and small jet aircraft fly in the troposphere within 6 km from the ground; Large and high speed jet aircraft are equipped with pressurization device, which can fly higher in the bottom of stratosphere, where almost no vertical airflow movement exits.

In the stratosphere, the aircraft flies smoothly, and the air is thin and the flight resistance is small. Therefore, the aircraft can fly at a higher speed. Most of the activities of modern civil aviation transportation are carried out in this layer. Ultrasonic speed aircraft and some high speed military aircraft cruise at 13.5–18 km or higher altitude in order to reduce resistance.

平流层位于对流层之上，顶层距地面约 50 km。

在平流层的下半部分（20 km 以下），平流层的温度几乎不随高度变化，年平均值为 −56.5 ℃。

飞机通常在平流层下部和对流层运行。无座舱增压装置的飞机和小型喷气式飞机一般在距离地面 6 km 内的对流层中飞行；大型和高速喷气式飞机配备了增压装置，可以在几乎没有垂直气流运动的平流层底部飞得更高。

在平流层，飞机飞行平稳，空气稀薄，飞行阻力小。因此，飞机可以以更高的速度飞行。现代民航运输的大部分活动都是在这一层进行的。超声速飞机和一些高速军用飞机在 13.5~18 km 或更高的高度巡航，以减少阻力。

 **New Words**

altitude	['æltɪtjuːd]	n.	海拔、海拔高度、高程
troposphere	['trɒpəsfɪə(r)]	n.	对流层（大气的最低层，在地球表面的 6~10 km 上空之间）
stratosphere	['strætəsfɪə(r)]	n.	平流层
latitude	['lætɪtjuːd]	n.	纬度、纬度地区
kilometer	['kɪləʊˌmiːtə]	n.	千米
equator	[ɪ'kweɪtə(r)]	n.	赤道
pole	[pəʊl]	n.	地极、磁极、电极
gravity	['grævəti]	n.	重力、地球引力
mass	[mæs]	n.	大量、许多
concentrate	['kɒnsntreɪt]	v.	集合、聚集
hail	[heɪl]	n.	雹、冰雹
phenomena	[fə'nɒmɪnə]	n.	现象（phenomenon 的复数）
horizontal	[ˌhɒrɪ'zɒntl]	adj.	水平的、与地面平行的、横的
vertical	['vɜːtɪkl]	adj.	竖的、垂直的、直立的、纵向的
airflow	[eə(r) fləʊ]		空气、气流
gust	[gʌst]	n.	强风、一阵狂风
physical parameter	['fɪzɪkl pə'ræmɪtə(r)]		物理参数
pressure	['preʃə(r)]	n.	压力、压强、大气压
density	['densəti]	n.	密集、稠密、密度
temperature	['temprətʃə(r)]	n.	温度、气温
sound speed	[saʊnd spiːd]		声速
aircraft	['eəkrɒft]	n.	飞机、航空器
cabin	['kæbɪn]	n.	（飞机的）座舱
pressurization	[ˌpreʃəraɪ'zeɪʃn]	n.	加压、增压
jet	[dʒet]	n.	喷气式飞机、喷射流、喷射口、喷嘴
resistance	[rɪ'zɪstəns]	n.	抵抗力、阻力、电阻
civil aviation	['sɪvl ˌeɪvi'eɪʃn]		民用航空
ultrasonic	[ˌʌltrə'sɒnɪk]	adj.	超声的
military	['mɪlətri]	adj.	军事的
cruise	[kruːz]	n.	巡航
altitude	['æltɪtjuːd]	n.	海拔、海拔高度、高程

Q&A

The following questions are for you to answer to assess the learning outcomes.

(1) Understand the composition of the atmosphere layers.

(2) What are the two layers of atmosphere from bottom to above?

(3) List the layers in which most of our aircraft are flying, and explain their position relations.

(4) Understand the physical properties of the atmosphere at different altitudes.

(5) What is the mean thickness of troposphere?

(6) What are the main characteristics of troposphere in terms of aircraft flying?

(7) What are the main characteristics of stratosphere in terms of aircraft flying?

(8) Analyze the relevant aerodynamics problems by applying the layers properties of the atmosphere.

(9) What basic requirements should an aircraft meet when flying at the altitude of stratosphere?

(10) List the advantages of flying at the altitude of stratosphere.

(11) In the altitude of troposphere, what influences of atmosphere should we consider to ensure the aircraft's safety and effectiveness?

 Extended Reading

Atmosphere

The Earth's atmosphere is so much more than the air we breathe. A trip from the surface of the Earth to outer space would result in passing through five different layers, each with very different characteristics.

Look up. Way up. The clouds you see in the sky, the wind that is moving the trees or the flag in your school yard, even the sunshine you feel on your face—these are all a result of the Earth's atmosphere.

The Earth's atmosphere stretches from the surface of the planet up to as far as 10,000 km above. After that, the atmosphere blends into space. Not all scientists agree where the actual upper boundary of the atmosphere is, but they can agree that the bulk of the atmosphere is located close to the Earth's surface–up to a distance of around 8–15 km.

While oxygen is necessary for most lives on the Earth, the majority of the Earth's atmosphere is not oxygen. The Earth's atmosphere is composed of about 78% nitrogen, 21% oxygen, 0.9% argon, and 0.1% other gases. Trace amounts of carbon dioxide, methane, water vapor, and neon are some of the other gases that make up the remaining 0.1%.

The atmosphere is divided into five different layers, based on temperature. The layer closest to the Earth's surface is the troposphere, reaching from about 7–15 km from the surface. The troposphere is the thickest at the equator, and much thinner at the North and South Poles. The majority of the mass of the entire atmosphere is contained in the troposphere approximately 75%–80%. Most of the water vapor in the atmosphere, along with dust and ash particles, are found in the troposphere—explaining why most of the Earth's clouds are located in this layer. Temperatures in the troposphere decrease with altitude.

The stratosphere is the next layer up from the Earth's surface. It reaches from the top of the troposphere, which is called the tropopause, to an altitude of approximately 50 km. Temperatures in the stratosphere increase with altitude. A high concentration of ozone, a molecule composed of three atoms of oxygen, makes up the ozone layer of the stratosphere. This ozone absorbs some of the incoming solar radiation, shielding life on the Earth from potentially harmful ultraviolet light, and is responsible for the temperature increase in altitude.

The top of the stratosphere is called the stratopause. Above that is the mesosphere, which reaches as far as about 85 km above the Earth's surface. Temperatures decrease in the mesosphere with altitude. In fact, the coldest temperature in the atmosphere is near the top of the mesosphere—about −90 ℃. The atmosphere is thin here, but still thick enough so that meteors will burn up as they pass through the mesosphere—creating what we see as "shooting stars". The upper boundary of the mesosphere is called the mesopause.

The thermosphere is located above the mesopause and reaches out to around 600 km. Not much is known about the thermosphere except that temperatures increase with altitude. Solar radiation makes the upper regions of the thermosphere very hot, reaching a temperature as high as 2,000 ℃.

The uppermost layer, that blends with what is considered to be outer space, is the exosphere. The pull of the Earth's gravity is so small that molecules of gas escape into outer space.

任务 3　大气的性质
Task 3　Properties of the Atmosphere

Contents

1) Density

2) Temperature

3) Pressure

4) Relative humidity

5) Viscosity

6) Compressibility of the air

7) Speed of sound

⊙ Learning Outcomes

1) Master the definitions of the atmosphere properties

2) Master the principles of the atmosphere properties

3) Analyze the change principles of the atmosphere properties

4) Solve the aerodynamics problems by the properties of the atmosphere

5) Cultivate professional qualities of rigor, carefulness, and ability to express, coordinate, and communicate effectively

▤ 任务内容

1）密度

2）温度

3）大气压

4）相对湿度

5）黏度

6）大气可压缩性

7）声速

⊙ 任务目标

1）掌握大气的物理参数定义

2）掌握大气物理参数的意义

3）分析大气物理参数的变化规律

4）运用大气物理参数性质解决相关的空气动力学问题

5）培养严谨、细心的职业素养，以及有效表达、协调和沟通的能力

Learning Guide

The flight environment includes various factors and weather phenomena, which can have an impact on the structure, onboard equipment, flight guidance trajectory, flight performance and controllability, stability, etc. of aircraft.

 课文 1

Density

密度

Atmospheric density refers to the air mass per unit volume. It decreases with the increase of altitude and changes approximately according to an exponential curve（Fig. 1–3）.

At about 22,000 ft (1ft=0.304,8 m), the atmospheric density is about half of that at sea level.

Properties of the Atmosphere (1)

大气密度是指单位体积的空气质量。它随着海拔的升高而减小，并大致按照指数曲线变化（图 1–3）。

在约 22 000 ft（1ft=0.304 8 m）处，大气密度约为海平面的一半。

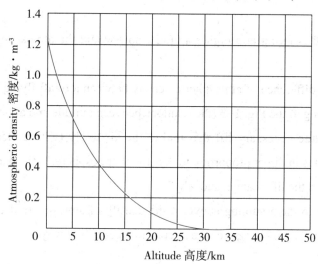

Fig. 1–3　Altitude vs. Density
图 1–3　高度与大气密度的关系

 New Words

curve	[kɜːv]	*n.*	曲线、弧线、曲面、弯曲
feet	[fiːt]	*n.*	英尺（单位）
meter	['miːtə(r)]	*n.*	米（单位）
sea level	[siː 'levl]		海平面
proportion	[prə'pɔːʃn]	*n.*	倍数关系、比例

decrease	[dɪ'kriːs]	v.	（使大小、数量等）减少、减小、降低
weight	[weɪt]	n.	重量、重物
lift	[lɪft]	n.	升力
runway	['rʌnweɪ]	n.	飞机跑道
takeoff	['teɪ,kɔf]	n.	（飞机的）起飞

 Q&A

The following questions are for you to answer to assess the learning outcomes.

(1) Identify and write down the definitions of the atmospheric density.

(2) Say the principles of the atmospheric density.

(3) What factors would effect the atmospheric density?

(4) Analyze the change principles of the atmospheric density.

(5) Analyze the reason for change of atmospheric density.

(6) Analyze the relevant aerodynamics problems by applying the density properties of the atmosphere.

(7) If an aircraft is flying at a higher altitude, how would the atmospheric density change, and why?

(8) Explain the difference of atmospheric density between aircraft flying at 5 km and 10 km respectively, according to the Fig. 1–3 (the relations between altitude and atmospheric density).

(9) Assuming: the necessary lift that makes the air to fly (which means the lift should contours the gravity) is in direct proportion to density of atmosphere. If the density of atmosphere decreases, how would the lift change, and why?

(10) According to the assuming above, if the density of atmosphere decreases, how would the necessary runway length change for the aircraft to takeoff, and why?

 Extended Reading

What Is the Speed of Sound?

1. What Is a Sound Wave?

First, let me talk about just sound in the air. Of course, you can have sound waves underwater (hello submarines) and even through solids. But think of the air. At one level, air is made up of a whole bunch of tiny particles. Oh sure, it is really more complicated than just tiny air particles. It is mostly nitrogen gas with some oxygen. But in this model of sound waves, it is fine to think of them all as just small particles (Fig. 1–4).

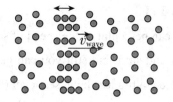

Fig. 1-4　Sound wave
图 1-4　声波

What happens if you take a whole bunch of these particles and push them all at the same time? Well, the pushed particles will go a little ways, but they will collide with other air particles and push them. Those particles will collide with more and so on and so on. This is what we call a wave. The important thing to realize is that the air isn't moving very far, but the compression is moving.

Another great example of this is the wave in a football stadium. Here is an example in case you have no idea what I am talking about.

What goes around the stadium? The people? No, they just move up and down. It is the disturbance that moves as the wave. The same is true for sound waves in the air. OK, but that is just a simple model of sound. How fast is the sound wave in the air? Although 340 m/s is a good first answer, it isn't always true. Let's look back the wave of people in a football stadium. What would make this speed change? Two things could clearly make a difference. Suppose the stadium wasn't full but instead about every other seat was occupied. This could change the speed of the sound wave. It isn't quite clear if it would make it faster or slower, but I would guess faster since the person would be reacting to the previous person that was farther away. Another effect could be from the alertness level of the crowd. If people weren't paying much attention, it could cause a longer reaction time and thus a lower wave speed.

Actually, now I am curious. I wonder if stadium wave speeds are fairly constant for different stadiums and crowds. My guess is that they would all have similar speed values. Maybe this will be a future blog post.

OK, back to sound waves in the air. What does this speed depend on? You could guess a few things. Just like the football crowd wave, the density of particles should matter. And what about the pressure in the air? That should matter too, right? Surprisingly (at least for me), a simple model for the speed of sound only varies with air temperature. Why? Well, as you get higher in the altitude (up to a point), the temperature decreases. The pressure and the density of air also decrease. The effects due to pressure and density essentially negate each other. Like I said, this over simplifies the whole issue. The Wikipedia page on the speed of sound has a lot more details if you are interested.

2. Altitude vs. Speed of Sound

If you put this together, you can get a plot of the speed of sound as a function of altitude. Oh sure, it will change with weather and stuff, but still you can get a pretty basic model. Here is a plot of the speed of sound at different altitudes above sea level (Fig. 1-5).

At sea level, the value is right around the 340 m/s mark. If you move up to 120,000 ft, the speed will drop down to around 200 m/s. Just from this data, you can see that Felix Baumgartner did indeed fall faster than the speed of sound. However, the question doesn't really make sense. Did he fall faster than the speed of sound at sea level? Yes. Was he also going faster than the speed of sound for the altitude he was at? Well, it makes logical sense that if the speed of sound is the greatest at sea level and he went faster than the speed of sound he would be going faster than the local speed of sound.

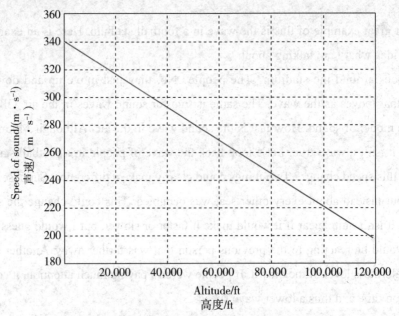

Fig. 1-5　Speed of sound changes with altitude

图 1-5　高度与声速的关系

3. Speed vs. Local Speed of Sound

I don't know if "local speed of sound" is an official term, but I like it. I am using it to mean the speed of sound at the current altitude. Here is a plot of the speed of Felix as he falls along with the plot of the local speed of sound at that same time(Fig. 1-6).

You will notice that from this numerical calculation, Felix was going faster than the local speed of sound for about 45 s. You should also notice that this calculation has his maximum speed a little over the reported value of 373 m/s—hopefully I can fix this later when I compare my model to the real data—but it's not too far off.

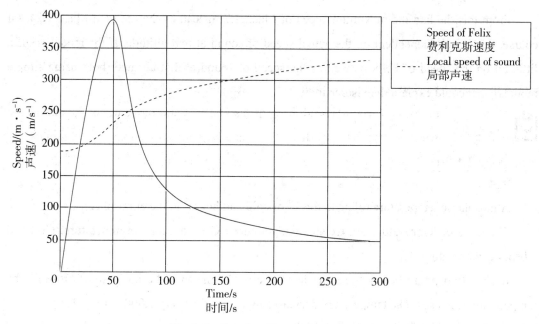

Fig. 1-6   Speed vs. Local speed of sound
图 1-6   速度与局部声速

4. Mach Number

I guess I was right (at least according to Wikipedia). It has the definition of Mach number(Ma) as the ratio of the speed of an object to the local speed of sound. Here is a plot (Fig. 1-7) of the speed of Felix as a function of altitude in terms of the Mach number (again, this is based on my not so perfect model).

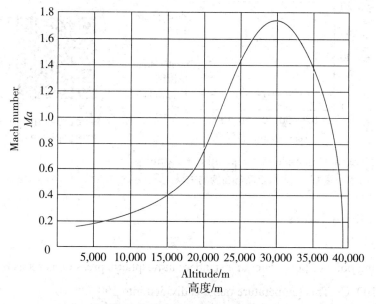

Fig. 1-7   Mach number Altitude
图 1-7   马赫数与高度的关系

From this, he had the maximum speed of Mach 1.7 instead of the reported Mach 1.24. Of course, this is very dependent on the actual speed of sound at that altitude. If the model is off a little bit on the speed of Felix as well as the speed of sound at that altitude (both using simple models), that could explain the discrepancy.

 课文 2

Temperature

温度

Atmospheric temperature refers to the ℃ of cold and hot air in the atmosphere.

The units of temperature are Celsius temperature (℃), Fahrenheit temperatures (℉) and Absolute temperature (K).

In the atmosphere below the altitude of about 11 km, with the increase of altitude, the temperature decreases. The temperature decreases by 6.5 ℃ for every 1,000 m (Fig. 1–8).

大气温度是指大气中空气冷热的程度。

温度单位为摄氏温度（℃）、华氏温度（℉）和绝对温度（K）。

在海拔 11 km 以下的大气中，随着海拔的升高，温度降低。海拔每升高 1 000 m 温度下降 6.5 ℃（图 1-8 所示）。

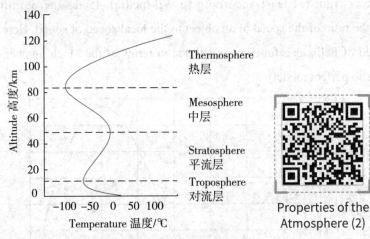

Properties of the Atmosphere (2)

Fig. 1–8 Atmospheric temperature vs. Altitude

图 1-8 大气温度与高度的关系

1. Celsius Temperature

1. 摄氏温度

The freezing point of pure water at a standard atmospheric pressure is set as 0 ℃, the boiling point is set as 100 ℃. This temperature range is divided into 100 ℃.

纯水在标准大气压下的冰点设定为 0 ℃，沸点设定为 100 ℃。该温度范围分为 100 ℃。

2. Fahrenheit Temperature

2. 华氏温度

The freezing point of pure water at a standard atmospheric pressure is set as 32 ℉, and the temperature of boiling water is set to 212 ℉. The temperature range between freezing and boiling points of water is divided into 180 ℉.

纯水在标准大气压下的冰点设定为 32 ℉，沸水温度为 212 ℉。水的冰点和沸点之间的温度范围划分为 180 ℉。

3. Absolute Temperature

3. 绝对温度

Absolute temperature is the temperature when the speed of molecules is stationary, its temperature interval is the same as that of centigrade.

绝对温度是分子速度不变时的温度，其温度区间与摄氏度相同。

4. Conversion of Three Temperature Units

4. 三种温度单位的换算

Conversion of Celsius, Fahrenheit and Absolute temperature units is as follows.

$$℃ =(℉-32) \, 5/9$$

$$K=273.15+℃$$

摄氏、华氏和绝对温度单位的换算如下。

$$℃ =(℉-32)5/9$$

$$K=273.15+℃$$

 New Words

Celsius	['selsiəs]	*n.* 摄氏
		adj. 摄氏温度的
Fahrenheit	['færənhaɪt]	*n.* 华氏
		adj. 华氏温度的
freezing point	['friːzɪŋ pɔɪnt]	冰点
molecule	['mɒlɪkjuːl]	*n.* 分子
unit	['juːnɪt]	*n.* 单位、单元
formula	['fɔːmjələ]	*n.* 公式、方程式、计算式

 Q&A

The following questions are for you to answer to assess the learning outcomes.

(1) Identify the definitions of the atmospheric temperature and the units of temperature.

(2) Write down the definition of the atmospheric temperature.

(3) Write down the definition of the Celsius temperature.

(4) Write down the definition of the Fahrenheit temperature.

(5) Write down the definition of the Absolute temperature.

(6) Write down the relation formulas of the three types of temperature units.

(7) Say the principles of the atmospheric temperature.

(8) Analyze the change principles of the atmospheric temperature.

(9) What factors would effect the change of the atmospheric temperature?

(10) In the troposphere, how does the temperature change with altitude?

(11) Analyze the relevant aerodynamics problems by applying the temperature of the atmosphere.

(12) Assumption: When the temperature increases, the density of the atmosphere would decrease. If the temperature of atmosphere decreases, how would the necessary runway length change for the aircraft to take off, and why?

(13) What are relations of the temperatures at the altitude of 5 km and 10 km?

(14) Assumption: The temperature at sea level is 25 ℃ . What is the temperature at 8 km altitude?

 课文 3

1. Atmospheric Pressure

1. 大气压

Atmospheric pressure refers to the pressure of the air in the atmosphere, that is, a force in an area pushed vertically against a unit area of an object (Fig. 1–9).

Properties of the Atmosphere (3)

At certain temperatures, the pressure decreases with the increase of the altitude (Fig. 1–10).

Field pressure, baro–corrected pressure and standard sea level pressure are commonly used in aviation. The field pressure is the pressure of the highest point in the landing airport. The baro–corrected pressure is calculated by the field pressure according to the atmospheric conditions and the flying altitude. The standard sea level pressure is the value of sea level pressure under standard atmospheric conditions (Fig. 1–11).

The commonly used units of air pressure are Pascal (Pa) and millimeter of mercury (mmHg), psi (Pounds per square inch, 1 inch=2.54 cm) are often used in aviation. Conversion of Pa, mmHg and psi is as below.

$$1\ \text{mmHg} = 133\ \text{Pa}$$

$$1\ \text{psi} = 6\ 895\ \text{Pa}$$

大气压是指大气中的空气压强，即空气对物体单位面积的垂直作用力（图 1-9）。在一定温度下，压力随高度的增加而降低（图 1-10）。

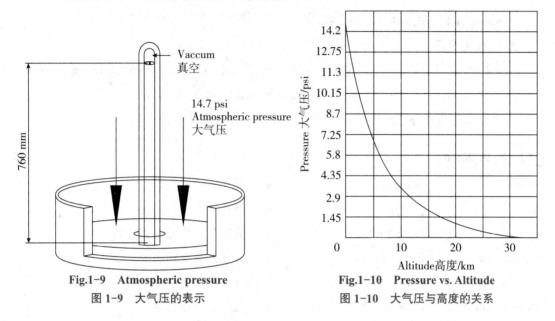

Fig.1-9　**Atmospheric pressure**
图 1-9　大气压的表示

Fig.1-10　**Pressure vs. Altitude**
图 1-10　大气压与高度的关系

场面气压、修正海平面气压和标准海平面气压通常用于航空领域。场面气压是着陆区域最高点的气压。根据大气条件、飞行高度以场压计算修正海平面气压。标准海平面气压是标准大气条件下海平面的气压（图 1-11）。

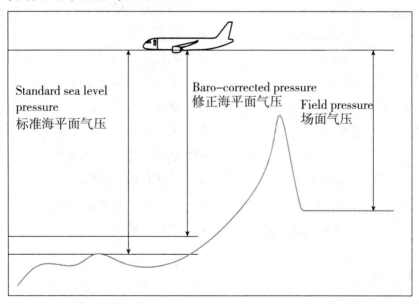

Fig. 1-11　**Field pressure, baro-corrected pressure and standard sea level pressure**
图 1-11　场面气压、修正海平面气压、标准海平面气压

常用的气压单位为帕斯卡（Pa）和毫米汞柱（mmHg），航空中通常使用磅/平方英寸 (psi)。Pa、mmHg 和 psi 之间的关系如下。

$$1 \text{ mmHg} = 133 \text{ Pa}$$

$$1 \text{ psi} = 6\,895 \text{ Pa}$$

2. Standard Atmospheric Pressure
2. 标准大气压

Since atmospheric pressure varies with the altitude and the temperature, the standard atmospheric pressure is defined as the pressure at the sea level in 15 ℃ .

大气压随海拔和温度变化，海平面 15 ℃的气压被定义为标准大气压。

 New Words

pound	[paʊnd]	n.	磅（质量单位）
inch	[ɪntʃ]	n.	英寸（长度单位）
square	[skweə(r)]	n.	正方形、平方、二次幂
vacuum	['vækjuːm]	n.	真空、真空状态
field pressure	[fiːld 'preʃə(r)]		场面气压
baro-corrected	[bɒrəʊ kə'rektɪd]		修正气压、气压修正
aviation	[ˌeɪvi'eɪʃn]	n.	航空
landing	['lændɪŋ]	n.	降落、着陆
		v.	降落、着陆
airport	['eəpɔːt]	n.	航空港、机场
calculate	['kælkjuleɪt]	v.	计算、推测
value	['væljuː]	n.	数值、作用
Pascal	['pæskl]	n.	帕斯卡（压强单位）
millimeter	['mɪlɪˌmiːtə]	n.	毫米（长度单位）
mercury	['mɜːkjəri]	n.	汞、水银
psi			磅/平方英寸（压强单位）
cabin	['kæbɪn]	n.	（飞机的）座舱
pilot	['paɪlət]	n.	飞行员、驾驶员、领航员
passenger	['pæsɪndʒə(r)]	n.	乘客、旅客

 Q&A

The following questions are for you to answer to assess the learning outcomes.

(1) Identify and write down the definition of the atmospheric pressure.

(2) List the most commonly used pressure units in aviation.

(3) Write down the relation formula of the pressure units in aviation.

(4) Write down the definition of the field pressure.

(5) Write down the definition of the baro−corrected pressure.

(6) Write down the definition of the standard sea level pressure.

(7) Identify the principles of the atmospheric pressure.

(8) Analyze the change principles of the atmospheric pressure.

(9) Explain the change principles of the atmospheric pressure with the altitude.

(10) Explain the relations of the atmospheric pressure at the altitude of 5 km and 10 km.

(11) Analyze the relevant aerodynamics problems by applying the pressure of the atmosphere.

(12) Assumption: The aircraft is flying at an altitude of 10 km, the pressure inside the cabin is the same as the ambient pressure. What dangerous incidents would happen to the pilots and passengers?

(13) Assumption: The airport is at an altitude of 5 km. How would the field pressure change, compared to an airport at the sea level?

 课文 4

Relative Humidity
相对湿度

Properties of the
Atmosphere (4)

Relative humidity refers to the ratio of the amount of water vapor in the atmosphere, to the maximum amount of water vapor that can be contained in the atmosphere at the same temperature. When the relative humidity is 100%, the amount of water vapor has reached the limit, and the water vapor is saturated.

Under different temperatures, the maximum amount of water vapor in the atmosphere is different. The higher the temperature, the greater water vapor it can contain. Therefore, with the decrease of temperature, the relative humidity would increase.

The temperature at which the relative humidity reaches 100% is called dew point temperature. At this temperature, the water vapor in the atmosphere has reached saturation state and began to condense, resulting in various meteorological phenomena such as cloud and fog.

There are some special weather phenomena in the high−altitude areas due to the condensation of the water vapor. For example, when a jet plane is flying at a high altitude, there will be one or several long cloud belts behind the plane. This is a special cloud formed by the mixture of exhausted gas by the aircraft and the surrounding air (Fig. 1−12).

The density of the water vapor is lighter than the one of the air, so the humid air is lighter than the dry air, which would affect the flight performance of the aircraft.

相对湿度是指大气中的水蒸气量与相同温度下大气中可包含的最大水蒸气量之比。当相对湿度为 100% 时，水蒸气量已达到极限，水蒸气已饱和。

在不同的温度下，大气中的最大水蒸气量是不同的：温度越高，它所能包含的水蒸气越多。因此，随着温度的降低，相对湿度会增加。

相对湿度达到 100% 的温度称为露点温度。在这个温度下，大气中的水蒸气已达到饱和状态并开始凝结，从而形成云和雾等各种气象现象。

由于水汽凝结，在高海拔地区有一些特殊的天气现象。例如，当喷气式飞机在高空飞行时，飞机后面会有一条或几条长云带。这是一种特殊的云，它由飞机排出的废气和周围空气混合而成（图 1-12）。

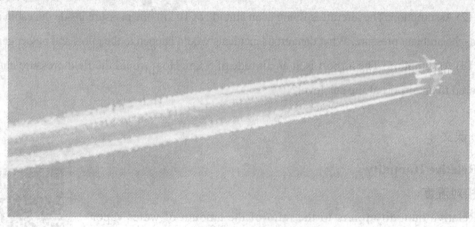

Fig. 1-12　Clouds behind the plane
图 1-12　飞机尾迹

水蒸气的密度比空气轻，因此潮湿空气比干燥空气轻，这将影响飞机的飞行性能。

 New Words

humidity	[hjuː'mɪdəti]	n.	（空气中的）湿度
relative	['relətɪv]	adj.	比较的、相对的
ratio	['reɪʃiəʊ]	n.	比率、比例
maximum	['mæksɪməm]	adj.	最多的、最大的
saturate	['sætʃəreɪt]	v.	浸透、使饱和
dew point	[djuː pɔɪnt]		露点（温度）
state	[steɪt]	n.	状态、情况
condense	[kən'dens]	v.	（气体）冷凝、（气体）凝结
fog	[fɒg]	n.	雾
mixture	['mɪkstʃə(r)]	n.	混合、混合物

exhaust	[ɪgˈzɔːst]	v.	耗尽、排气
performance	[pəˈfɔːməns]	n.	表现、性能
influence	[ˈɪnfluəns]	n.	影响、作用
skin	[skɪn]	n.	（飞机的）蒙皮
surface	[ˈsɜːfɪs]	n.	（飞机的）表面、舵面

 Q&A

The following questions are for you to answer to assess the learning outcomes.

(1) Identify and write down the definition of the atmospheric humidity.

(2) Write down the definition of the atmospheric relative humidity.

(3) Write down the definition of the atmospheric dew point.

(4) Identify the principles of the atmospheric humidity.

(5) Analyze the change principles of the atmospheric humidity.

(6) Explain the change principles of the atmospheric relative humidity with the temperature.

(7) Analyze the relevant aerodynamics problems by applying the humidity properties of the atmosphere.

(8) Assumption: When the air temperature increases, how would the relative humidity vary?

(9) Why can we find long cloud belts behind the plane?

(10) When the temperature increases, it will cause the air humidity to increase as well. If the temperature of atmosphere increases, how would the necessary runway length change for the aircraft to take off, and why?

(11) List some influences on the skin and surfaces of the aircraft due to the change of humidity.

 课文 5

Viscosity
黏度

Properties of the Atmosphere (5)

When the motion of two adjacent flow layers in a fluid is different, there would be a mutual drag occurring between the two layers. It refers to the viscosity of the fluid.

The viscosity force of the fluid is proportional to the motion difference between adjacent flow layers, to the contact area, and inversely proportional to the distance between flow layers.

Different fluids have different viscosity coefficients.

Atmospheric viscosity is relatively small, but its influence on aircraft flight can not be ignored.

A fluid without viscosity is called an ideal fluid. When the viscous effect is not considered, the air can be regarded as an ideal fluid.

当流体中两个相邻流动层的运动不同时，两层之间会产生相互阻力，即流体的黏度。

流体的黏性力与相邻流动层之间的运动差、接触面积成正比，与流动层之间距离成反比。

不同的流体具有不同的黏度系数。

大气的黏度相对较小，但其对飞机飞行的影响不容忽视。

没有黏度的流体称为理想流体。当不考虑黏性效应时，可以将空气视为理想流体。

 New Words

motion	['məʊʃn]	n.	运动、移动
fluid	['fluːɪd]	n.	液体、流体
mutual	['mjuːtʃuəl]	adj.	相互的
drag	[dræg]	n.	（流体的）阻力
viscosity	[vɪ'skɒsəti]	n.	（流体的）黏性、黏度
force	[fɔːs]	n.	作用力、力
proportional	[prə'pɔːʃənl]	adj.	成比例的
inversely	[ˌɪn'vɜːsli]	adv.	相反地、反向地
distance	['dɪstəns]	n.	距离、间距
coefficient	[ˌkəʊɪ'fɪʃnt]	n.	系数
ideal fluid	[aɪ'diːəl 'fluːɪd]		理想流体
viscous effect	['vɪskəs ɪ'fekt]		黏性效应

 Q&A

The following questions are for you to answer to assess the learning outcomes.

(1) Identify and write down the definition of the viscosity of fluid.

(2) How does the viscosity vary in different media?

(3) How does the viscosity vary with different temperature?

(4) What is the main influence of the viscosity on aircraft and flight activities?

(5) What kind of fluid can be regarded as ideal fluid?

(6) What is the purpose of the simplification of ideal fluid?

(7) List some phenomen a of the fluid viscosity in your life.

课文 6

Properties of the
Atmosphere (6)

1. Compressibility of the Air
1. 空气的可压缩性

Compressibility of the air refers to the change of volume and density of the air when the pressure or temperature changes.

Due to the large distance and the small attraction between air molecules, air has great compressibility.

When flying at a low speed, the change of the air pressure and density is small. The compressibility has little effect on the aircraft. In this condition, the compressibility can be ignored and the atmosphere can be regarded as an incompressible fluid.

When flying at a high speed, the change of the air pressure and density is great. The influence of the air compressibility on the flight can not be ignored. At this time, the compressibility of atmosphere must be considered.

空气的可压缩性是指当压力或温度变化时，空气体积和密度的变化。

由于分子之间的距离大，吸引力小，空气具有很高的压缩性。

当低速飞行时，空气压力和密度的变化很小。压缩性对飞机的飞行几乎没有影响。在这种情况下，可忽略可压缩性，将大气视为不可压缩流体。

当在高速飞行时，由于速度高，空气压力和密度变化很大。空气压缩性对飞行的影响不容忽视。此时，必须考虑大气的可压缩性。

2. Speed of Sound
2. 声速

Speed of sound is the propagation speed of weak pressare disturbance in the medium. It is an instance for air compressibility.

It is affected by the temperature and density of the atmosphere.

As the temperature increases, the speed of sound increases. In the same medium, the speed of sound is only related to the temperature.

$$a=20.1\sqrt{T}$$

The speed of sound (a) is proportional to the square root of the absolute temperature (T) of the airflow.

As the density of the propagating medium increases, the speed of sound increases.

声速是介质中微弱压强扰动的传播速度。它是空气可压缩性的示例。

Properties of the
Atmosphere (7)

声速受大气温度和密度的影响。

温度增加，声速增加。在同一种介质中，声速只与温度有关。

$$a=20.1\sqrt{T}$$

同一气体的当地声速 a 不是固定不变的，它与气体的绝对温度 T 的平方根成正比。

当传播介质的密度增加，声速也增加。

 New Words

compressibility	[kəm,prɛsɪ'bɪlɪti]	n.	可压缩性、压缩系数
attraction	[ə'trækʃn]	n.	吸引
sound speed	[saʊnd spiːd]		声速
propagation	[,prɒpə'geɪʃ(ə)n]	n.	（声音、振动）传播
disturbance	[dɪ'stɜːbəns]	n.	扰动、干扰
medium	['miːdiəm]	n.	（传播的）媒介、手段

 Q&A

The following questions are for you to answer to assess the learning outcomes.

(1) Identify and write down the definition of the compressibility of the air.

(2) Identify and write down the definition of the speed of sound.

(3) When the compressibility of the air is considered, will the flight speed of the aircraft increase or decrease?

(4) Assumption: Aircraft A is flying at a higher altitude than aircraft B, their flight speed to the ground are exactly the same. What are the differences of the air density at their flight altitudes?

(5) Assumption: Aircraft A is flying at a higher altitude than aircraft B, their flight speed to the ground are exactly the same. What are the differences of sound speed at their flight altitudes?

02
模块 2
Module 2

大气基本原理
Principles of Atmospheric

Contents

1) Principle of relative motion
2) Airflow field and airflow streamline
3) Air continuity equation
4) Bernoulli equation

学习内容

1）相对运动原理
2）气流场和气流流线
3）空气连续性方程
4）伯努利方程

任务 1 相对运动原理
Task 1 Principle of Relative Motion

Contents

Principle of relative motion

Learning Outcomes

1) Understand the principle of relative motion
2) Solve the aerodynamics problems by the principle of relative motion

3) Cultivate professional qualities of rigor, carefulness, and ability to express, coordinate, and communicate effectively

 任务内容

　　相对运动原理

 任务目标

　　1）理解相对运动原理
　　2）运用相对运动原理解决相关的空气动力学问题
　　3）培养严谨、细心的职业素养，以及有效表达、协调和沟通的能力

Learning Guide

　　Relative motion is familiar to everyone, and it is related to the choice of the reference. For example, when sitting on a train, we naturally choose the environment of the train as the reference to observe the surrounding things. When the train moves, the scenery outside the train is moving backwards, which means that the scenery moves backwards relative to the train. Of course, if the ground is taken as the reference, the observer on the ground will feel that the scenery on the bottom is stationary, as the train moves forward. This is the difference in the selection of reference frames, leading to different conclusions on relative motion.

 课文

Principle of Relative Motion
相对运动原理

The aerodynamic force acting on the aircraft depends on the relative motion between the aircraft and the air. In other words, when the aircraft moves in stationary air at a certain speed, the interaction force between the aircraft and the air are equivalent to the case when the aircraft is stationary and the air flows in the opposite direction. This is the principle of relativity.

The movement of the air relative to the aircraft is called relative wind (also known as relative flow incoming air flow or coming airflow in this book). The direction of relative airflow is opposite to the direction of the aircraft movement, as shown in Fig. 2–1. Since the relative air velocity is equivalent, the generated aerodynamic force has the same effect.

Thus, the flight movement is transformed into air flow movement, which greatly simplifies the analysis of aerodynamic problems.

Wind tunnel experiment is based on this principle.

作用在飞机上的气动力取决于飞机与空气之间的相对运动。换言之，当飞机以一定速度在静止的空气中移动时，飞机与空气之间的相互作用力等同于飞机静止且空气反向流动的情况。这就是相对运动原理。

空气相对于飞机的运动称为相对气流（来流）。相对气流的方向与飞机运动的方向相反，如图 2-1 所示。由于相对气流速度相等，因此产生的空气动力具有相同的效果。

因此，飞行运动被转化为气流运动，这大大简化了空气动力学问题的分析。

风洞实验就是基于这一原理。

Principle of
Relative Motion

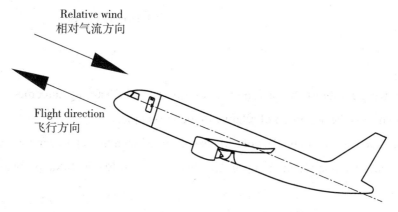

Fig. 2-1 The direction of flight and airflow
图 2-1 飞行方向和气流方向

New Words

relative	['relətɪv]	*n.*	亲戚，亲属，同类事物
		adj.	相对的，比较的
relativity	[ˌrelə'tɪvəti]	*n.*	相对论，相对性
motion	['məʊʃn]	*n.*	运动，移动
		v.	（以头或手）做动作，示意
interaction	[ˌɪntə'rækʃən]	*n.*	相互作用，相互影响
equivalent	[ɪ'kwɪvələnt]	*n.*	相等的东西，等量，对应词
		adj.	相等的，相同的
opposite	['ɒpəzɪt]	*n.*	对立的人，对立面，反面
		adj.	对面的，另一边的，相反的

		adv.	在对面
		prep.	与……相对，在……对面
direction	[dəˈrekʃn]	*n.*	方向，方位
principle	[ˈprɪnsəpl]	*n.*	规范，法则，原则，原理
generate	[ˈdʒenəreɪt]	*vt.*	生成，产生，引起
transform	[trænsˈfɔːrm]	*n.*	变换
		vt.	使改变，使转换
simplify	[ˈsɪmplɪfaɪ]	*vt.*	简化，使简易
analysis	[əˈnæləsɪs]	*n.*	分析，分析结果
wind tunnel	[ˈwɪnd tʌnl]		风洞，风道
experiment	[ɪkˈsperɪmənt]	*n.*	实验，试验
		vi.	做试验，进行实验

 Q&A

The following questions are for you to answer to assess the learning outcomes.

(1) Write downs the definition of relative motion principle.

(2) What practical experience in your life is concerned with the relative motion principle?

(3) How does the relative motion principle help us to resolve the basic problem of aviation study?

 Extended Reading

What Are Wind Tunnels?

Wind tunnels are large tubes with air moving inside. The tunnels are used to copy the actions of an object in flight. Researchers use wind tunnels to learn more about how an aircraft will fly. NASA uses wind tunnels to test scale models of aircraft and spacecraft. Some wind tunnels are big enough to hold full-size versions of vehicles. The wind tunnel moves air around an object, making it seem like the object is really flying.

1. How do Wind Tunnels Work?

Most of the time, powerful fans move the air through the tube. The object to be tested is fastened in the tunnel so that it will not move. The object can be a small model of a vehicle. It can be just a piece of a vehicle. It can be a full-size aircraft or spacecraft. It can even be a common object like a tennis ball (Fig. 2-2). The air moving around the still object shows what would happen if the object were moving through the air. How the air moves can be studied in different ways. Smoke or dye can be placed in the air and can be seen as it moves. Threads can be attached to the object to show how the air is moving. Special instruments are often used to measure the force of the air on the object.

Fig. 2-2　A wind tunnel test shows how a tennis ball moves through the air
图 2-2　风洞试验展示网球是如何在空气中运动的

2. How Does NASA Use Wind Tunnels for Aircraft?

NASA has more wind tunnels than any other group. The agency uses the wind tunnels in a lot of ways. One of the main ways NASA uses wind tunnels is to learn more about airplanes and how things move through the air. One of NASA's jobs is to improve air transportation. Wind tunnels help NASA test ideas for ways to make aircraft better and safer. Engineers can test new materials or shapes for airplane parts. Then, before flying a new airplane, NASA will test it in a wind tunnel to make sure it will fly as it should.

NASA also works with others that need to use wind tunnels. That way, companies that are building new airplanes can test how the planes will fly (Fig. 2-3). By letting these companies use the wind tunnels, NASA helps to make air travel safer.

Fig. 2-3　Airplane builders use NASA wind tunnels to test new airplane designs
图 2-3　飞机制造商使用 NASA 的风洞来测试新的飞机设计

3. How Can Wind Tunnels Help Spacecraft?

NASA also uses wind tunnels to test spacecraft and rockets. These vehicles are made to operate in space. Space has no atmosphere. Spacecraft and rockets have to travel through the atmosphere to get to space. Vehicles that take humans into space also must come back through the atmosphere to the Earth.

Wind tunnels are important. NASA uses them to test the Orion spacecraft and the Space Launch System rockets. These rockets are called the SLS. Orion and SLS are new vehicles. They will take astronauts into space. NASA must test the systems in wind tunnels to see if they are safe to fly, and NASA must see what happens when Orion comes back to the Earth through the atmosphere.

Wind tunnels can even help engineers design spacecraft to work on other worlds. Mars has a thin atmosphere. It is important to know what the Martian atmosphere will do to vehicles that are landing there. Spacecraft designs and parachutes are tested in wind tunnels set up to be like the Martian atmosphere.

4. What Types of Wind Tunnels Does NASA Use?

NASA has many different types of wind tunnels. They are located at NASA centers all around the country. The wind tunnels come in a lot of sizes. Some are only a few square inches (1 square inch=6.45 cm), and some are large enough to test a full−size airplane. Some wind tunnels test aircraft at very slow speeds. But some wind tunnels are made to test at hypersonic speeds. That is more than 4,000 miles (1mile=1 609.344 m) per hour!

任务 2　流场和流线
Task 2　Flow Field and Streamline

 Contents

    1) Flow field

    2) Unsteady flow and steady flow

    3) Streamline

    4) Diagram of streamlines

    5) Properties of streamlines

    6) Flow tube

Learning Outcomes

1) Master the concepts of flow field and streamline

2) Understand the concept and characteristic of steady and unsteady flow

3) Analyze the characteristic of streamline

4) Understand the concept of flow tube

5) Calculate the volume flow and mass flow of the flow tube

6) Cultivate professional qualities of rigor, carefulness, and ability to express, coordinate, and communicate effectively

任务内容

1）流场

2）非定常流和定常流

3）流线

4）流线谱

5）流线的特点

6）流管

任务目标

1）掌握流场和流线的概念

2）理解和掌握定常流体与非定常流体的概念和特点

3）分析流线的特点

4）理解流管的概念

5）计算流管的体积流量和质量流量

6）培养严谨、细心的职业素养，以及有效表达、协调和沟通的能力

Learning Guide

Flow field refers to a flow field where velocity, pressure, and other factors change. In flight, it is caused by the motion of the aircraft; In wind tunnel experiments, it is due to the model placed in the uniform straight airflow.

课文

Flow Field

流场

The space where fluid flows is called flow field.

流体运动所占的空间称为流场。

1. Steady and Unsteady Flow

1. 非定常流和定常流

The flow's parameters such as velocity, pressure, temperature and density, at any point in the flow field, are not fixed, variant with time. This flow is designated as unsteady flow, the corresponding flow field is an unsteady flow field.

The flow's parameters, such as velocity, pressure, temperature and density, at any point in the flow field, are fixed, invariant with time. This flow is designated as steady flow, the corresponding flow field is a steady flow field.

In the steady flow field, the flow parameters in a certain position are independent of time.

如果流场中任何一点的气流参数（速度、压力、温度和密度）随时间变化，则气流被指定为非定常流，这种流场即为非定常流场。

如果流场中任何一点的气流参数（速度、压力、温度和密度）随时间保持不变，则气流被指定为定常流，气流场即为定常流场。

在定常流场中，某一位置的流动参数与时间无关。

2. Streamline

2. 流线

The streamline is a spatial curve in the flow field. At each point on the streamline, the tangent direction of the curve designates the direction of the flow velocity of the fluid flowing through the point.

流线是流场中的空间曲线。在流线的每个点上，曲线的切线方向表示流经该点流体的流速方向。

3. Diagram of Streamlines

3. 流线谱

In the flow field, a picture depicting the flow of fluid clusters is defined as the diagram of streamlines (Fig. 2-4).

在流场中，用流线组成的描绘流体微团流动情况的图称为流线谱（图2-4）。

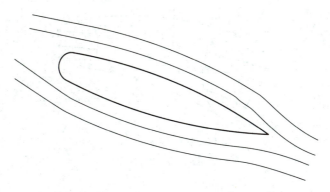

Fig. 2-4   Streamline and diagram of streamlines
图 2-4   流线和流线谱

4. Properties of Streamline

4. 流线的特点

Each point on the streamline has only one direction of motion.

Streamlines cannot cross or diverge.

The diagram of streamlines in steady flow field is time-invariant.

流线上每个点只有一个运动方向。

流线不能交叉或分叉。

如果流线谱不随时间变化，则为定常流。

5. Flow Tube

5. 流管

Take a closed curve that is not a streamline in the flow field. A tubular surface is formed by the streamline at each point on the curve, and the surface is a flow tube （Fig. 2-5）. The fluid flows only through the cross section of the flow tube, it does not flow in or out through the tube wall.

Flow Field and Streamline (1)

Assume the cross-sectional area of the flow tube is A, the fluid density is ρ, the flow velocity on the cross section is v, then the volume of fluid flowing through the section per unit time is Av, which is the volume flow of fluid. The fluid mass is ρAv, which is the mass flow of the fluid. The unit of fluid mass is kg/s.

在流场中任取不是流线的闭合曲线，通过曲线上每个点的流线形成管状表面，称为流管（图 2-5）。流体仅流经流管的横截面，不会通过管壁流入或流出。

Flow Field and Streamline (2)

假设流管的截面积为 A，流体密度为 ρ，截面上的流速为 v，则单位时间内流过该截面的流体体积为 Av，即流体的体积流量。流体质量为 ρAv，即流体的质量流量，其单位为 kg/s。

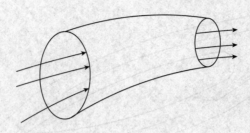

Fig. 2-5　Flow tube

图 2-5　流管

 New Words

flow field	[fləʊ fiːld]	流场；流线谱
streamline	[ˈstriːmlaɪn]	*n.* 流线型，流线
		vt. 使成流线型
		adj. 流线型的
spatial	[ˈspeɪʃl]	*adj.* 空间的
curve	[kɜːrv]	*n.* 曲线，弧线
		v. （使）沿曲线运动，呈曲线形
tangent direction	[ˈtændʒənt dəˈrekʃn]	切（线方）向
designate	[ˈdezɪgneɪt]	*vt.* 命名，指定，选定
		adj. 未上任的
velocity	[vəˈlɒsəti]	*n.* 速度，高速，快速
vary	[ˈveri]	*v.* 变化
unsteady flow	[ʌnˈstedi fləʊ]	非稳流
steady flow	[ˈstedi fləʊ]	定常流动，稳定流动
independent	[ˌɪndɪˈpendənt]	*n.* 独立
		adj. 独立的，自主的
cross	[krɔːs]	*n.* 十字形记号
		v. 交叉，穿越，越过
		adj. 恼怒的，十分愤怒的，生气的
diverge	[daɪˈvɜːrdʒ]	*vi.* 偏离，分歧
diagram of streamlines	[ˈdaɪəgræm əv ˈstriːmlaɪnz]	流线谱
depict	[dɪˈpɪkt]	*vt.* 描写，描述
closed	[kləʊzd]	*v.* 关闭，关
		adj. 封闭的
tubular surface	[ˈtuːbjələr ˈsɜːrfɪs]	管状曲面

flow tube	[fləʊ tuːb]		流管
cross section	['krɔːs sekʃn]	n.	横截面（图）；剖面（图）
area	['eriə]	n.	部位；领域；面积
volume	['vɒljuːm]	n.	体积；容积；容量
unit time	['juːnɪt taɪm]		单位时间
volume flow	['vɒljuːm fləʊ]		体积流量
mass flow	[mæs fləʊ]		质量流量

 Q&A

The following questions are for you to answer to assess the learning outcomes.

(1) Describe the definition of the flow field.

(2) Write down the definition of the streamline.

(3) How can the flow field and flow streamline help to study the principles of airflow?

(4) What are the characteristics of the streamline in the field?

(5) Write down the definition of the flow tube.

(6) What are the characteristics of the steady and unsteady flow?

(7) Can you list some examples of steady and unsteady flow in your life?

任务 3 空气连续性方程
Task 3 Air Continuity Equation

 Contents

    1) Air continuity hypothesis

    2) Air continuity equation

Learning Outcomes

    1) Master air continuity equation

    2) Solve the aerodynamic problems by air continuity equation

    3) Cultivate professional qualities of rigor, carefulness, and ability to express, coordinate, and communicate effectively

任务内容

1）空气连续性假设

2）空气连续性方程

任务目标

1）掌握空气连续性方程的表达

2）运用连续性方程解决相关的空气动力学问题

3）培养严谨、细心的职业素养，以及有效表达、协调和沟通的能力

Learning Guide

We often say the law of conservation of mass, but did you know that the law of conservation of mass is called the continuity theorem in aerodynamics. Let's talk about the continuity theorem.

课文

Air Continuity Hypothesis
空气连续性假设

The air is composed of molecules, there are gaps between molecules, and the molecules constantly move randomly. The average distance between two consecutive collisions is called the average travelling distance of molecules. Because the length of the aircraft (such as the distance between the two wingtips) is often greater than the average travelling distance of the air molecules, the micro characteristics of the air can be ignored and the macro ones are considered. That is the air continuity hypothesis, which treats the air as a continuous fluid without gap.

In the aerodynamic analysis, the atmosphere is regarded as a continuous medium according to the continuity hypothesis. The air cluster is composed of countless molecules, and the characteristics of the cluster are the common ones of the molecules.

空气是由分子组成的，分子之间有间隙，分子不断地随机移动。两次连续碰撞之间的平均距离称为分子的平均行程。由于飞机的长度（如两个翼尖之间的距离）通常大于空气分子的平均行程，因而可以忽略空气的微观结构，只考虑其宏观特性。这就是空气连续性假说，它将空气视为无间隙的连续流体。

在空气动力学分析中，连续性假设是将大气视为连续介质，其中任何微空气团由无数分子组成，微空气团的特性反映了许多分子的共同特性。

1. Air Continuity Equation
1. 空气连续性方程

The air continuity equation is an application of law of conservation of mass in steady flow. Assuming the air is flowing through the wing at speed v in steady flight, there are three cross sections named 1, 2, and 3 in the flow tube. The mass flow of the cross sections is:

$$q_{m1} = \rho_1 A_1 v_1, \; q_{m2} = \rho_2 A_2 v_2 \cdots$$

According to the law of conservation of mass:

$$q_{m1} = q_{m2} = \cdots$$

In the steady flow, the fluid flows continuously and stably in the flow tube, and the mass flow through each section of the flow tube is constant.

$$\rho_1 A_1 v_1 = \rho_2 A_2 v_2 \cdots$$

When flying at a low speed, the atmosphere can be simplified as an incompressible fluid, which means the density is constant. The continuity equation can be simplified as:

$$A_1 v_1 = A_2 v_2 = \cdots$$

连续性方程是质量守恒定律在流体定常流动中的应用。假设空气在稳定飞行中以速度 v 流经机翼，流管中有三个截面，分别命名为 1、2 和 3。这些截面的质量流量为

$$q_{m1} = \rho_1 A_1 v_1, \; q_{m2} = \rho_2 A_2 v_2 = \cdots$$

根据质量守恒定律：

$$q_{m1} = q_{m2} = \cdots$$

在定常流中，流体在流管中连续稳定地流动，流经流管各部分的质量流量恒定。

$$\rho_1 A_1 v_1 = \rho_2 A_2 v_2 = \cdots$$

当以低速飞行时，可以认为空气是不可压缩流体，即密度恒定。连续性方程可以简化为

$$A_1 v_1 = A_2 v_2 = \cdots$$

2. Properties of Air Continuity Equation
2. 空气连续性方程的性质

The speed of the fluid is inversely proportional to the cross-sectional area of the flow tube: The thinner the flow tube is, the faster the flow velocity is. The flow tube becomes wider, the streamline becomes sparse, and the flow velocity becomes slower.

流体的速度与流动管的横截面积成反比：流管越细，流体速度越快；流管变粗，流线变稀疏，流速变慢。

Air Continuity
Equation (1)

Air Continuity
Equation (2)

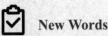 **New Words**

continuity	[ˌkɒntɪˈnuːəti]	*n.*	连续性，持续性
molecules	[ˈmɒləˌkjulz]	*n.*	分子
gap	[gæp]	*n.*	间隙，间隔，开口
		vi.	造成缝隙，使成缺口
constantly	[ˈkɒnstəntli]	*adv.*	不断地，一直，始终
randomly	[ˈrændəmli]	*adv.*	随机地，随意地
average	[ˈævərɪdʒ]	*n.*	平均数，平均水平
		v.	平均为，计算出……的平均数
		adj.	平均的
consecutive	[kənˈsekjətɪv]	*adj.*	连续的，连续不断的
collisions	[kəˈlɪʒənz]	*n.*	碰撞
wingtip	[ˈwɪŋˌtɪp]		翼尖；翼梢
ignore	[ɪgˈnɔːr]	*vt.*	忽视
micro	[ˈmaɪkrəʊ]	*n.*	微型
structure	[ˈstrʌktʃər]	*n.*	结构，构造
		vt.	使形成体系
macro	[ˈmækrəʊ]	*n.*	宏指令
		adj.	巨大的
characteristic	[ˌkærəktəˈrɪstɪk]	*n.*	特征，特点
		adj.	特有的，典型的
continuous	[kənˈtɪnjuəs]	*adj.*	不断的，持续的
medium	[ˈmiːdiəm]	*n.*	形式，介质
		adj.	中等的，中号的
cluster	[ˈklʌstər]	*n.*	簇，团，束
		vi.	群聚，聚集
countless	[ˈkaʊntləs]	*adj.*	无数的
equation	[ɪˈkweɪʒn]	*n.*	方程式，方程，等式
airfoil	[ˈerfɔɪl]	*n.*	翼型
conservation of mass	[ˌkɒnsəˈveɪʃ(ə)n əv mæs]		质量守恒
stably	[ˈsteibli]	*adv.*	稳定地，坚固地
mass	[mæs]	*n.*	团，块
		adj.	大批的，数量极多的
		v.	集结，聚集

incompressible	[ɪnkəm'prɛsəbəl]	*adj.* 不能压缩的；坚硬的
inversely	[ˌɪn'vɜːslɪ]	*adj.* 相反地；倒转地
proportional to	[prə'pɔːʃən(ə)l tuː]	正比于，与……成比例
thin	[θɪn]	*n.* 细小部分
		v. 使稀薄，使变淡
		adv. 薄薄地
		adj. 薄的，细的，瘦的

 Q&A

The following questions are for you to answer to assess the learning outcomes.

(1) Describe the air continuity equation.

(2) Please explain the physical principle of the air continuity equation.

(3) In what case is the volume flow equal to the mass flow?

(4) Please list some examples that contains the fundamental of the air continuity equation in your life.

任务 4　伯努利定理
Task 4　Bernoulli Theorem

 Contents

1) Aerodynamic force

2) Bernoulli equation

3) Static pressure energy

4) Dynamic energy

5) Properties of Bernoulli equation

 Learning Outcomes

1) Master Bernoulli equation

2) Solve the aerodynamic problems by Bernoulli equation

3) Cultivate professional qualities of rigor, carefulness, and ability to express,

coordinate, and communicate effectively

 任务内容

1) 空气动力
2) 伯努利方程
3) 压力能
4) 动力能
5) 伯努利方程的性质

任务目标

1) 掌握伯努利方程的表达
2) 运用伯努利方程解决相关的空气动力学问题
3) 培养严谨、细心的职业素养，以及有效表达、协调和沟通的能力

 Learning Guide

Bernoulli theorem is widely used in hydraulics and applied fluid mechanics. The Bernoulli effect is applicable to all ideal fluids, including liquids and gases, and is one of the basic phenomena of stable fluid flow. It reflects the relationship between fluid pressure and flow velocity, as well as the relationship between flow velocity and pressure.

$$\frac{1}{2}\rho v^2 + P = P_0$$

课文

Aerodynamic Force
空气动力

Aircraft are vehicles which are able to fly by gaining support from the air. The gravity is counteracted by either the static lift (Fig. 2-6) or the dynamic lift of an airfoil. In a few cases, the downward thrust from jet engines is adopted as well.

A fluid flowing past the surface of a body exerts a surface force on it. If the fluid is air, the force is called an aerodynamic force.

飞机是通过获得空中支持而能够飞行的交通工具。它通过使用机翼的静态升力（图 2-6）或动态升力，或在少数情况下使用喷气发动机的向下推力来抵消重力。

流经物体表面的流体对物体施加表面力。如果流体是空气，这种力称为空气动力。

Fig. 2-6　Airship static lift
图 2-6　飞艇的静态升力

1. Bernoulli Equation

1. 伯努利方程

Bernoulli equation is the application of the law of energy conservation in fluid flow.

The law of energy conservation means that in an isolated system, energy can be converted from one form to another, but the sum of energy remains unchanged.

Bernoulli equation states that, when an incompressible and ideal fluid flows steadily and slowly in a system without energy exchange to the outside, the energy form in the system can be converted to each other, but the total energy remains unchanged.

In the above case, the energy of fluid flow is only converted between static pressure energy, dynamic energy and gravitational potential energy.

The Bernoulli principles can be formulated as:

$$\begin{cases} P_s + P_d = P_0, \ P_0 = \text{const} \\ P_d = \frac{1}{2}\rho v^2 \\ P_s + \frac{1}{2}\rho v^2 = P_0, \ P_0 = \text{const} \end{cases}$$

where P_s is the static pressure, P_d is the dynamic pressure, and P_0 is the total pressure.

伯努利方程是能量守恒定律在流体流动中的应用。

能量守恒定律意味着在孤立系统中，能量可以从一种形式转换为另一种形式，但能量之和保持不变。

伯努利方程指出，当不可压缩的理想流体在系统中稳定缓慢地流动，而没有能量交换到外部时，系统中的能量形式可以相互转换，但总能量保持不变。

在上述情况下，流体流动的能量仅在压力能、动能和重力势能之间转换。

伯努利原理可以表述为：

$$\begin{cases} P_s + P_d = P_0, \ P_0 = \text{const} \\ P_d = \frac{1}{2}\rho v^2 \\ P_s + \frac{1}{2}\rho v^2 = P_0, \ P_0 = \text{const} \end{cases}$$

式中，P_s 是静压；P_d 是动压；P_0 是总压。

2. Static Pressure Energy

2. 压力能

Static pressure energy is the work capacity of a fluid due to pressure. It is designated as P_s, pressure energy per unit volume of fluid. In stationary air, the static pressure is atmospheric pressure.

压力能是流体因压力而产生。P_s 为单位体积流体的压力能。在静止空气中，静压是大气压力。

3. Dynamic Energy

3. 动力能

Dynamic energy is the ability of a fluid to do work due to its velocity. It is designated as P_d,

$$P_d=\frac{1}{2}\rho v^2$$

where ρ is the density of the fluid, v is the velocity of the fluid.

动力能是流体因其速度而做功的能力。

$$P_d=\frac{1}{2}\rho v^2$$

式中，ρ 为流体密度；v 为流体速度。

4. Properties of Bernoulli Equation

4. 伯努利方程的性质

When the fluid flows with little height variation, it can be considered that the gravitational potential energy of the fluid remains unchanged. In this way, there is only the mutual conversion between static pressure energy and dynamic energy in the flow.

The total pressure is the sum of static pressure and dynamic pressure.

When an incompressible and ideal fluid flows steadily, the flow tube becomes thinner, the streamline becomes denser, the flow speed of the fluid will increase, the dynamic pressure of the fluid will increase, and the static pressure will decrease.

When the flow tube becomes wider, the streamline becomes more sparse, the flow speed of the fluid will decrease, the dynamic pressure of the fluid will decrease and the static pressure will increase.

When the air flows through the pipe continuously and steady at a certain speed, as shown in Fig. 2-7, at the thinnest section of the pipe, the speed is the fastest and the pressure of the air drops the most. At the widest section of the pipe, the speed is the slowest, the pressure is the highest.

When the air flows through the wing surface, the sections of the flow tube formed on the wing surface varies with the direction of the air flow and the airfoil used by the wing. Thus, the static pressure energy and dynamic energy in the fluid change, and different pressure distributions are formed on the wing surface, resulting in generating the lift (Fig. 2-8).

　　当流体在高度变化不大的情况下流动时，可以认为流体的重力势能保持不变。这样，流体中只有压力能和动力能相互转换。

　　总压力是静态压力和动态压力之和。

　　当不可压缩的理想流体稳定流动时，流管变细，流线变密，流体流速增加，流体动态压力增加，静态压力降低。

　　当流管变粗时，流线变稀疏，流体流速降低，流体动态压力降低，静态压力升高。

　　当空气以一定速度连续稳定地流过管道时，如图 2-7 所示，在管道最窄的部分，速度最快，空气压力下降最大。在管道最宽的部分，速度最慢，压力最高。

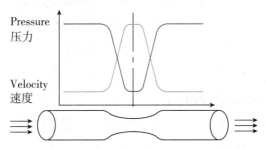

Fig. 2-7　The air speed and pressure
图 2-7　空气流动的速度和压力

　　当空气流经机翼表面时，在机翼表面形成的流管截面随气流方向和机翼使用的翼型而变化。因此，流体中的压力能和动力能发生变化，机翼表面形成不同的压力分布，从而产生升力（图 2-8）。

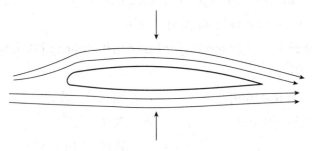

Fig. 2-8　Generation of lift
图 2-8　升力的产生

Bernoulli
Equation (1)

Bernoulli
Equation (2)

Bernoulli
Equation (3)

Bernoulli
Equation (4)

New Words

dynamic lift	[daɪ'næmɪk lɪft]		动态升力
static lift	['stætɪk lɪft]		静态升力；浮力
gravity	['grævəti]	*n.*	重力，地球引力
vehicles	['viːəhɪkəlz]	*n.*	交通工具，车辆
downward	['daʊnwərd]	*adj.*	向下的，下降的
		adv.	向下地，下降地
jet	[dʒet]	*n.*	喷气式飞机，喷射流，喷射口
		v.	乘坐喷气式飞机
		adj.	喷气式推进的
thrust	[θrʌst]	*v.*	挤，推
		n.	推力，驱动力
aerodynamic	[ˌerəʊdaɪ'næmɪk]	*adj.*	空气动力学的
conservation of energy	[ˌkɒnsə'veɪʃ(ə)n əv 'enədʒi]		能量守恒
isolate	['aɪsəleɪt]	*n.*	分离
		vt.	隔离，脱离
form	[fɔːrm]	*n.*	类型，种类，形式
sum	[sʌm]	*v.*	概括，总结
energy	['enərdʒi]	*n.*	能量，能源
pressure energy	['preʃər 'enərdʒi]		压力能，静压能
dynamic energy	[daɪ'næmɪk 'enərdʒi]		动能
gravitational potential energy	[ˌgrævɪ'teɪʃənl pə'tenʃl 'enərdʒi]		重力势能
height	[haɪt]	*n.*	高，高度
mutual	['mjuːtʃuəl]	*adj.*	相互的，彼此的
convert	[kən'vɜːrt]	*n.*	改变
		v.	转换，(使)转变
conversion	[kən'vɜːrʒn]	*n.*	转变，转换，转化
capacity	[kə'pæsəti]	*n.*	容量，容积，电容
static pressure	['stætɪk 'preʃər]		静态压力
dynamic pressure	[daɪ'næmɪk 'preʃər]		动态压力
dense	[dens]	*adj.*	稠密的，密集的
distribution	[ˌdɪstrɪ'bjuːʃn]	*n.*	分配，分布
lift	[lɪft]	*n.*	升力
		v.	举起，抬高

The following questions are for you to answer to assess the learning outcomes.

(3) Bernoulli principles of gas can be regarded as the application of what principle in the process of air flow?

(4) Can you explain the flight principle of an airplane?

 Extended Reading

The Principle and Applications of Bernoulli Equation

Bernoulli's principle is an important theory in fluid mechanics, involves a lot of knowledge of fluid mechanics, and was proposed by Daniel Bernoulli (Swiss physicist, mathematician, 1700—1782) in 1726, is the basic equation of the three hydrodynamics . It is the embodiment of objects mechanical energy conversion of hydraulics.

Agricultural sprayer is a kind of this machine. How can a sprayer spray water or liquid into a fog? As shown in Fig. 2-9, push the horizontal tube, then the air is pushed through a narrow place by the Bernoulli principle. The narrow place causes a larger flow rate and low pressure, There is a vertical tube CB connected with the narrow part of the horizontal tube, in which the pressure is less than the bottom pressure above the liquid. Under positive pressure and negative pressure, the liquid is ejected from the nozzle.

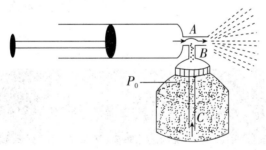

Fig. 2-9　Sprayer for agriculture
图 2-9　农用喷雾器

机翼形状和参数
Wing Airfoil and Parameters

Contents

1) Wing airfoil
2) Wing planform (which is the shape and layout of wing)
3) Wing installation parameters

学习内容

1）机翼的翼型
2）机翼的平面形状
3）机翼的安装参数

任务 1　机翼的翼型
Task 1　Wing Airfoil

 Contents

1) Definition of airfoil
2) Chord line
3) Thickness
4) Mean camber line
5) Relative camber

Learning Outcomes

1) Master the definition of geometric parameters of wing airfoil

2) Analyze the shape characteristics of the wing airfoil by the geometric parameters

3) Solve the aerodynamic problems by the geometric parameters

4) Cultivate professional qualities of rigor, carefulness, and ability to express, coordinate, and communicate effectively

任务内容

1）机翼翼型的定义

2）弦线

3）厚度

4）中弧线

5）相对弯度

任务目标

1）掌握机翼翼型几何参数的定义

2）运用机翼翼型几何参数分析机翼的形状特点

3）运用机翼翼型几何参数解决飞行中的相关空气动力学问题

4）培养严谨、细心的职业素养，以及有效表达、协调和沟通的能力

Learning Guide

Planes generally have symmetrical wings. The cross−section of the wing, which is cut parallel to the symmetrical plane of the wing, is called the wing profile and is usually also known as the airfoil. The geometric shape of an airfoil is one of the fundamental geometric characteristics of a wing. The aerodynamic characteristics of airfoils directly affect the aerodynamic characteristics of wings and the entire aircraft, and play an important role in aerodynamics theory and aircraft design.

课文

Definition of Airfoil
机翼翼型的定义
The shape of the cross−section of the wing is the wing airfoil. For straight wing, the wing airfoil can be observed by cutting the wing with a plane parallel to the symmetrical plane of the

fuselage (Fig. 3-1).

机翼的横截面形状为机翼翼型。对于直翼，机翼的翼型可以通过平行于机身对称平面的平面机翼剖面来观察（图 3-1）。

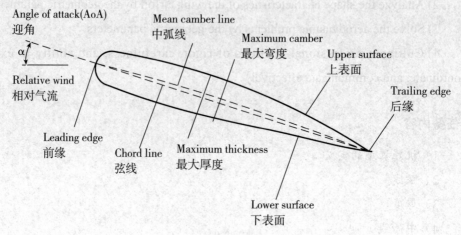

Fig. 3-1　Parameters of an airfoil
图 3-1　翼型参数

1. Chord Line
1. 弦线

The leading edge is the point at the front of the airfoil that has the maximum curvature, and the trailing edge is defined similarly as the point of the maximum curvature at the rear of the airfoil. The chord line is a straight line connecting the leading and trailing edges of the airfoil. The chord length, or simply chord, is the length of the chord line. The thickness of an airfoil varies along the chord.

前缘是翼型前部具有最大曲率的点。后缘被类似地定义为翼型后部的最大曲率点。弦线是连接机翼前缘和后缘的直线。弦长，简称弦，是弦线的长度。机翼的厚度沿弦线变化。

2. Thickness
2. 厚度

The distance between the upper and lower surfaces of the airfoil is the thickness. The maximum thickness is often at the forward part of the airfoil. The ratio of the maximum thickness to the chord length is the relative thickness, which also indicates the thickness of the airfoil. A small relative thickness indicates that the airfoil is thin. When the flight speed increases, the relative thickness decreases, and the position of the maximum thickness moves backward gradually. The relative thickness in civil transport aircraft is about 8%–16%, the position of the maximum thickness is about 35%–50%.

机翼上下表面之间的距离为厚度。最大厚度通常位于机翼的前部。最大厚度与弦长之比是相对厚度，表示机翼的厚度。较小的相对厚度表明翼型较薄。当飞行速度增加时，相对厚度减小，最大厚度位置逐渐向后移动。民用运输机的相对厚度为 8% ～ 16%，最大厚度位置为 35% ～ 50%。

3. Mean Camber Line

3. 中弧线

The mean camber line is the locus of points midway between the upper and lower surfaces. The shape of the airfoil is defined by using this concept.

中弧线是上下表面中间点的连线。翼型的形状是用这个概念定义的。

4. Relative Camber

4. 相对弯度

The maximum distance between the mean camber line and the chord is the maximum camber. The ratio of the maximum camber to chord length is relative camber. The relative camber indicates the camber degree of the airfoil. A large relative camber indicates a large camber degree, a small relative camber indicates a small degree of airfoil curvature. Low speed aircraft wing has large camber, and the maximum camber position is at the front of the airfoil.

中弧线和弦之间的最大距离为最大弯度。最大弯度与弦长之比为相对弯度。相对弯度表示翼型的弯曲程度，相对弯度大表示翼型弯曲程度大，相对弯度小表示翼型弯曲程度小。低速飞机的翼型具有较大的弯度，最大弯曲处位于翼型的前半部分。

5. Examples of Airfoil

5. 翼型举例

Different airfoils can be obtained by changing the parameters, such as camber, thickness, leading edge radius and trailing edge angle, as shown in Fig. 3-2.

Clark Y airfoil, whose upper surface is curved and lower surface is flat, is a typical subsonic airfoil. Laminar airfoil, with round leading edge and sharp trailing edge, is an asymmetric biconvex airfoil with camber. For symmetrical airfoil, the maximum camber is zero, the mean camber line coincides with the chord line, used in high speed aircraft to reduce the resistance.

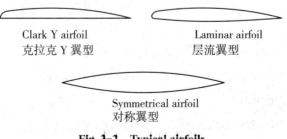

Clark Y airfoil
克拉克 Y 翼型

Laminar airfoil
层流翼型

Symmetrical airfoil
对称翼型

Fig. 3-2　Typical airfoils
图 3-2　常见翼型

不同的翼型可以通过改变参数来获得，如弯度、厚度、前缘半径和后缘角度，如图 3-2 所示。

克拉克 Y 翼型是一种典型的亚声速翼型，其上表面是弯曲的，下表面是平坦的。层流翼型具有圆形前缘和尖锐后缘，是一种不对称的双凸翼型。对称翼型，最大弯度为零，中弧线与弦线重合，用于高速飞机以减小阻力。

Wing Airfoil (1)   Wing Airfoil (2)   Wing Airfoil (3)

 New Words

straight	[streɪt]	*adj.*	直的，笔直的
parallel	[ˈpærəlel]	*adj.*	平行的
symmetrical plane	[sɪˈmetrɪkl pleɪn]		对称面
fuselage	[ˈfjuːsəlɑʒ]	*n.*	机身
chord line	[kɔːrd laɪn]		翼弦线，弦线
leading edge	[ˌliːdɪŋ ˈedʒ]		机翼前缘
trailing edge	[ˌtreɪlɪŋ ˈedʒ]		机翼后缘
rear	[rɪr]	*n.*	后部
		v.	抚养，养育
		adj.	后方的，后面的，后部的
thickness	[ˈθɪknəs]	*n.*	厚，厚度
upper	[ˈʌpər]	*n.*	上面
		adj.	上面的，上部的
lower	[ˈləʊər, ˈlaʊər]	*v.*	降低；减少
		adj.	下面的，下方的
		adv.	低，向下
maximum thickness	[ˈmæksɪməm ˈθɪknəs]		最大厚度
forward	[ˈfɔːrwərd]	*n.*	前部，前端
ratio	[ˈreɪʃiəʊ]	*n.*	比率，比例
relative thickness	[ˈrelətɪv ˈθɪknəs]		相对厚度
mean camber line	[miːn ˈkæmbər laɪn]		中弧线
locus	[ˈləʊkəs]	*n.*	中心，核心

radius	['reɪdɪəs]	n.	半径
angle	['æŋgl]	n.	角，斜角
		v.	斜移，斜置
flat	[flæt]	n.	公寓
		adv.	水平地，平整地
		adj.	平的，水平的
subsonic	[ˌsʌb'sɒnɪk]	n.	亚声速飞机
		adj.	亚声速的
laminar	['læmɪnər]	adj.	层状的，板状的
asymmetric	[ˌeɪsɪ'metrɪk]	adj.	不对称的，不对等的
biconvex	[baɪ'kɒnveks]	adj.	两面凸的，双凸的
symmetrical	[sɪ'metrɪkl]	adj.	对称的
coincide	[ˌkəʊɪn'saɪd]	vi.	重合，相符；重叠
resistance	[rɪ'zɪstəns]	n.	电阻，阻力
civil	['sɪvl]	adj.	民事的，民用的

 Q&A

The following questions are for you to answer to assess the learning outcomes.

(1) Please briefly describe the definition of wing airfoil.

(2) Please briefly describe the characteristics of subsonic airfoils.

(3) Please briefly describe the characteristics of supersonic airfoils.

(4) Under what circumstances are the mean camber line and the chord line overlapped into the same line?

 Extended Reading

Airfoil

The wings and stabilizers of fixed-wing aircraft, as well as helicopter rotor blades, are built with airfoil-shaped cross sections. Airfoils are also found in propellers, fans, compressors and turbines. Sails are also airfoils, and the underwater surfaces of sailboats, such as the centerboard, rudder, and keel, are similar in cross-section and operate on the same principles as airfoils. Swimming and flying creatures and even many plants and sessile organisms employ airfoils/hydrofoils. Common examples are bird wings, the bodies of fish, and the shape of sand dollars. An airfoil-shaped wing can create downforce on an automobile or other motor vehicles, improving traction.

When the wind is obstructed by an object such as a flat plate, a building, or the deck of a bridge, the object will experience drag and also an aerodynamic force perpendicular to the wind. This does not mean the object qualifies as an airfoil. Airfoils are highly-efficient lifting shapes, able to generate more lift than similarly sized flat plates of the same area, and able to generate lift with significantly less drag. Airfoils are used in the design of aircraft, propellers, rotor blades, wind turbines and other applications of aeronautical engineering.

The curve represents an airfoil with a positive camber, so some lift is produced at zero angle of attack. With increased angle of attack, lift increases in a roughly linear relation, called the slope of the lift curve. At about 18° this airfoil stalls, and lift falls off quickly over 18°. The drop in lift can be explained by the action of the upper-surface boundary layer, which separates and greatly thickens over the upper surface at and past the stall angle. The thickened boundary layer's displacement thickness changes the airfoil's effective shape, in particular it reduces its effective camber, which modifies the overall flow field so as to reduce the circulation and the lift. The thicker boundary layer also causes a large increase in pressure drag, so that the overall drag increases sharply near and past the stall point.

Airfoil design is a major facet of aerodynamics. Various airfoils serve different flight regimes. Asymmetric airfoils can generate lift at zero angle of attack, while a symmetric airfoil may better suit frequent inverted flight as in an aerobatic airplane. In the region of the ailerons and near a wingtip a symmetric airfoil can be used to increase the range of angles of attack to avoid spin-stall. Thus, a large range of angles can be used without boundary layer separation. Subsonic airfoils have a round leading edge, which is naturally insensitive to the angle of attack. The cross section is not strictly circular, however, the radius of curvature is increased before the wing achieves maximum thickness to minimize the chance of boundary layer separation. This elongates the wing and moves the point of maximum thickness back from the leading edge.

Supersonic airfoils are much more angular in shape and can have a very sharp leading edge, which is very sensitive to angle of attack. A supercritical airfoil has its maximum thickness close to the leading edge to have a lot of length to slowly shock the supersonic flow back to subsonic speeds. Generally, such transonic airfoils and also the supersonic airfoils have a low camber to reduce drag divergence. Modern aircraft wings may have different airfoil sections along the wing span, each one optimized for the conditions in each section of the wing.

Movable high-lift devices, flaps and sometimes slats, are fitted to airfoils on almost every aircraft. A trailing edge flap acts similarly to an aileron; however, it, as opposed to an aileron, can be retracted partially into the wing if not used.

A laminar flow wing has the maximum thickness in the middle camber line. Analyzing the

Navier–Stokes equations in the linear regime shows that a negative pressure gradient along the flow has the same effect as reducing the speed. So, with the maximum camber in the middle, maintaining a laminar flow over a larger percentage of the wing at a higher cruising speed is possible. However, some surface contamination will disrupt the laminar flow, making it turbulent. For example, with rain on the wing, the flow will be turbulent. Under certain conditions, insect debris on the wing will cause the loss of small regions of laminar flow as well. Before NASA's research in the 1970s and 1980s, the aircraft design community understood from application attempts in the WWⅡ era that laminar flow wing designs were not practical using common manufacturing tolerances and surface imperfections. That belief changed after new manufacturing methods were developed with composite materials (e.g. laminar–flow airfoils developed by Professor Franz Wortmann for use with wings made of fibre–reinforced plastic). Machined metal methods were also introduced. NASA's research in the 1980s revealed the practicality and usefulness of laminar flow wing designs and opened the way for laminar–flow applications on modern practical aircraft surfaces, from subsonic general aviation aircraft to transonic large transport aircraft, to supersonic designs.

Schemes have been devised to define airfoils—an example is the NACA system. Various airfoil generation systems are also used. An example of a general–purpose airfoil that finds wide application, and pre–dates the NACA system, is the Clark–Y. Today, airfoils can be designed for specific functions by the use of computer programs.

任务 2　机翼的平面形状
Task 2　Wing Planform

 Contents

    1) Definition of wing planform

    2) Wing area, taper ratio, wingspan, aspect ratio, sweepback

Learning Outcomes

    1) Master the definition of geometric parameters of wing planform

    2) Analyze the characteristics by geometric parameters of the wing planform

3) Solve the aerodynamic problems by parameters of the wing planform

4) Cultivate professional qualities of rigor, carefulness, and ability to express, coordinate, and communicate effectively

 任务内容

1）机翼平面形状的定义

2）机翼面积、梢根比、翼展、展弦比、后掠角

 任务目标

1）掌握机翼平面形状几何参数的定义

2）运用机翼平面形状几何参数分析机翼的形状特点

3）运用机翼平面形状几何参数解决飞行中的相关空气动力学问题

4）培养严谨、细心的职业素养，以及有效表达、协调和沟通的能力

Learning Guide

The palnform of the wing refers to the projected shape of the wing on the plane when viewed from the top of the aircraft.

课文

1. Definition of Wing Planform

1. 机翼形状的定义

When viewed from above, the projection shape of the wing is the wing planform, as shown in Fig. 3-3.

从上面看，机翼的投影形状就是机翼形状，如图 3-3 所示。

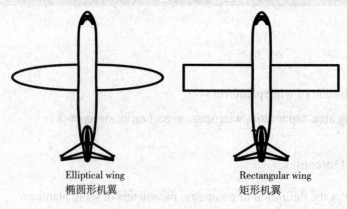

Elliptical wing
椭圆形机翼

Rectangular wing
矩形机翼

Fig. 3-3 Wing planform

图 3-3 机翼形状

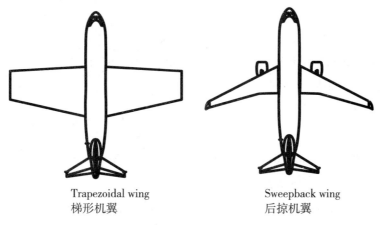

Trapezoidal wing
梯形机翼

Sweepback wing
后掠机翼

Fig. 3-3　Wing shapes (add)
图 3-3　机翼形状（续）

Geometric parameters of wings are shown in Fig. 3-4.

机翼的几何参数如图 3-4 所示。

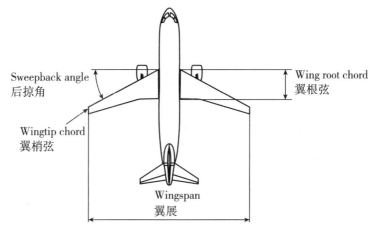

Sweepback angle
后掠角

Wing root chord
翼根弦

Wingtip chord
翼梢弦

Wingspan
翼展

Fig. 3-4　Wing parameters
图 3-4　机翼的几何参数

2. Wing Area

2. 机翼面积

The projected area of the wing is the wing area.

机翼的投影面积为机翼面积。

3. Taper Ratio

3. 梢根比

Ratio of the length of the tip chord to the one of the root chord is taper ratio.

翼梢弦长与根部弦长之比为梢根比。

4. Wingspan

4. 翼展

The distance from one wingtip to the other wingtip is wingspan.

左右翼梢之间的距离为翼展。

5. Aspect Ratio

5. 展弦比

The ratio of the wingspan to the average chord length, where average chord length is the average wingtip chord length and root chord length.

The wings with a large aspect ratio are often used in modern civil transport aircraft. With the increase of the flight speed, the aspect ratio will decrease.

展弦比是翼展与平均弦长之比，其中，平均弦长为翼尖弦长与根部弦长的平均值。

现代民用运输飞机通常使用大展弦比的机翼。随着飞行速度的增加，展弦比将减小。

6. Sweepback Angle

6. 后掠角

The angle between the line of equal percentage of the chord points along the span of the wing and the straight line perpendicular to the centerline of the fuselage. There are leading edge sweepback, 1/4 chord line sweepback.

Wing Planform

沿机翼翼展方向的等百分比弦线点的连线与垂直于机身中心线的直线之间的角度。有前缘后掠、1/4 弦线后掠。

📋 **New Words**

projection	[prə'dʒekʃn]	*n.*	投射，投影
taper ratio	['teɪpər 'reɪʃiəʊ]		梢根比
ratio	['reɪʃiəʊ]	*n.*	比率，比例
tip	[tɪp]	*n.*	翼尖，尖端
		v.	(使)倾斜，轻碰
root	[ruːt]	*n.*	翼根
		v.	(使)生根
wingspan	['wɪŋspæn]	*n.*	翼展，翼幅
aspect ratio	['æspekt reɪʃiəʊ]		纵横比，展弦比
sweepback	['swiːpbæk]	*n.*	后掠，后掠角
span	[spæn]	*n.*	跨度，范围
		vt.	跨越，横跨

percentage	[pər'sentɪdʒ]	*n.*	百分率，百分比
perpendicular	[ˌpɜːrpən'dɪkjələr]	*adj.*	垂直的，成直角的
		n.	垂直线
centerline	['sentəˌlaɪn]	*n.*	中心线

 Q&A

The following questions are for you to answer to assess the learning outcomes.

(1) How to judge whether the wing of an aircraft is a straight wing, a forward−swept wing or a backward−swept wing?

(2) Please briefly describe what the taper ratio is?

(3) Please briefly describe what the aspect ratio is?

 Extended Reading

What Are the Different Wing Planform?

The wing planform for each aircraft is mainly based on the aerodynamic requirements. There are other considerations like stealth, controllability, etc. The basic definition of the wing geometry is given in Fig. 3−5.

Most of the wing planforms in use (or have been used) fall under one of these categories.

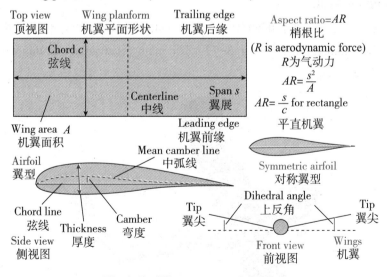

Fig. 3−5   Wing geometry definition
图 3−5   机翼几何定义

1. Elliptical Wing

Aerodynamically, the elliptical planform (Fig. 3−6) is the most efficient as elliptical spanwise lift distribution has the lowest possible induced drag (as given by thin airfoil theory). However, the most important disadvantage of the elliptical wing is that its manufacturability is poor.

Fig. 3-6　Elliptical wing

图 3-6　椭圆形机翼

Interestingly, the elliptical wing was not decided to minimize induced drag, but to house the retractable landing gear along with guns and ammunition inside a wing that had to be thin.

2. Rectangular Wing

Arguably the simplest wing planform from a manufacturing point of view, the rectangular wing is a straight, untapered wing（Fig. 3-7）. The main disadvantage of this wing is that it is aerodynamically inefficient.

Fig. 3-7　Rectangular wing

图 3-7　矩形机翼

3. Tapered Wing

This is a modification of the rectangular wing where the chord is varied across the span to approximate the elliptical lift distribution. While not as efficient as the elliptical lift distribution, it offers a compromise between manufacturability and efficiency (Fig. 3–8).

Fig. 3–8　Tapered wing
图 3–8　梯形机翼

4. Constant Chord with Tapered Outer

This is midway between the rectangular and the tapered wing, with the inner part having constant chord and the outer part having a taper (Fig. 3–9). Almost all of the wings seen so far are used in subsonic aircraft due to the large drag caused by them in the transonic regime.

Fig. 3–9　Constant chord with tapered outer wing
图 3–9　弦长固定，外侧为梯形的机翼

5. Delta Wing

The delta is a very low aspect ratio wing used in supersonic aircraft, most notably in the European designs (Fig. 3-10). The main advantages of the delta wing is that it is efficient in all the flight regimes (subsonic, transonic and supersonic). Also, the wing offers a large wing area for the shape, reducing wing loading and improving maneuverability.

Fig. 3-10 Delta wing
图 3-10 三角翼

The delta wing design is also very strong structurally, offering large volume for internal fuel. The delta wings are also quite simple to build and maintain.

The main disadvantages are that they have high induced drag due to low aspect ratio and also that they should have high angle of attack at low speeds (takeoff and landing), mainly due to fact that at these speeds, lift is generated by vortices. To compensate for this, they have high stall angles.

A variant of the delta wing used in some aircraft is the one with horizontal stabilizer (tailed delta) like Fig. 3-11 shows.

Fig. 3-11 Delta wing with horizontal stabilizer
图 3-11 带水平安定面的三角翼

Another variant is the one used by the Eurofighter Typhoon, the so-called cropped delta wing (Fig. 3-12), where the tips of the delta are "cut off" to reduce drag at high angles of attack.

Fig. 3-12　Cropped delta wing
图 3-12　翼尖钝化三角翼

Yet another variant of the delta wing (this planform is very popular in combat aircraft), is the so-called double delta (Fig. 3-13), where the leading edge angle is not constant, but has two different values.

Fig. 3-13　Double delta wing
图 3-13　复合三角翼

6. Trapezoidal Wing

The trapezoidal wing (Fig. 3-14) is a high performance configuration such that the leading edge sweeps back and the trailing edge sweeps forward. This is mostly found in combat aircraft from the US.

This wing configuration offers efficient supersonic flight and has very good stealth characteristics. However, the wing loading is quite high, resulting in reduced maneuverability, especially instantaneous turn rate.

Fig. 3-14　Trapezoidal wing
图 3-14　梯形机翼

7. Ogive Wing

The ogive is a type of supersonic wing used in high speed aircraft (Fig. 3-15). This is a complex mathematical shape derived for minimizing drag, especially at supersonic speeds. They offer excellent supersonic performance, with minimal drag. However, they are extremely complex and manufacturing is difficult while their subsonic performance is poor in comparison.

The wing planforms can also be classified by their sweep and variability.

Fig. 3-15　Ogive wing
图 3-15　子弹头型机翼

8. Swept Back Wing

The leading edges of these wings are swept back wings (Fig. 3-16). This is done in order to reduce drag in transonic speeds, which is determined by the velocity normal to the wing. This planform is used in almost all high speed commercial airliners.

Fig. 3-16   Swept back wing
图 3-16   后掠机翼

9. Swept Forward Wing

One disadvantage of the swept back wings (Fig. 3-17) is that due to the flow characteristics, the ailerons stall before the flaps (i.e. outboard wings stall first), which can lead to controllability issues. In order to overcome this, swept forward wings are used in a few (experimental) aircraft.

The main problem with swept forward wing is that it produces wing twisting as it bends under load, resulting in greater stress on the wing root than for a similar straight or swept back wing.

Fig. 3-17   Swept forward wing
图 3-17   前掠机翼

10. Variable Sweep Wings

For high speeds (transonic and supersonic), the swept wing is most suitable while for low speed (subsonic) flight, unswept wings are better. Variable sweep wings (Fig. 3-18) are used to optimize the wing planform over a wide range of speeds. The main problem with this type of wing is the mechanical complexity.

Fig. 3-18   Variable sweep wings

图 3-18   可变后掠机翼

任务 3   机翼的安装参数
Task 3   Wing Installation Parameters

 Contents

1) Monoplane & biplane

2) Low/mid/high wing aircraft

3) Dihedral, anhedral, incidence angle

4) Influence of installation parameters on aircraft performance

Learning Outcomes

1) Master the definition of the installation parameters

2) Analyze the characteristics of the wing by the installation parameters

3) Solve the aerodynamic problems by the installation parameters

4) Cultivate professional qualities of rigor, carefulness, and ability to express, coordinate, and communicate effectively

任务内容

1）单翼飞机和双翼飞机

2）上 / 中 / 下单翼飞机

3）上反角、下反角、纵向上反角

4）安装参数对飞机性能的影响

任务目标

1）掌握机翼安装参数的定义

2）运用机翼安装参数分析机翼的形状特点

3）运用机翼安装参数解决飞行中的相关空气动力学问题

4）培养严谨、细心的职业素养，以及有效表达、协调和沟通的能力

Learning Guide

According to the installation position and angle of the wing, there can be different installation parameters of the wings.

课文

Monoplane & Biplane
单翼飞机和双翼飞机

The wings of an aircraft produce lift. Many different styles and arrangements of wings have been used (Fig. 3-19). Most early fixed-wing aircraft were biplanes, whose wings were stacked one above the other. Nowadays, most aircrafts are monoplanes with one wing on each side.

飞机的机翼产生升力。不同样式和布局的机翼产生升力的方式不同（图 3-19）。最早的固定翼飞机是双翼飞机，机翼上下并列分布。现在大多数是单翼飞机，每侧有一个机翼。

Fig. 3-19　Biplane and monoplane
图 3-19　双翼机和单翼机

1. Low/Mid/High Wing Aircraft

1. 下 / 中 / 上单翼飞机

Fixed-wing aircraft are generally characterized by their wing configuration (Fig. 3-20). The aircraft whose wings are mounted on the lower fuselage is low wing aircraft. The aircraft whose wings are mounted approximately on the half way up the fuselage is mid wing aircraft. The aircraft whose wings are mounted on the upper fuselage is high wing aircraft.

固定翼飞机通常以其机翼配置为特征（图 3-20）。机翼安装在机身下部的飞机称为下单翼飞机；机翼安装在机身的一半左右的飞机称为中单翼飞机；机翼安装在机身上部的飞机称为上单翼飞机。

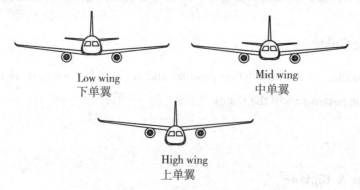

Low wing
下单翼

Mid wing
中单翼

High wing
上单翼

Fig. 3-20　Low/mid/high wing aircraft
图 3-20　下单翼、中单翼和上单翼

2. Dihedral, Anhedral, Incidence Angle

2. 上反角、下反角和纵向上反角

1) Dihedral Angle

1）上反角

The tips of the wing are higher than the roots (Fig. 3-21) as Boeing 737, giving a shallow "V" shape when viewed from the front, adds lateral stability.

尖端比根部高，如波音 737，从正面看呈浅 "V" 形，增加横向稳定性。

2) Anhedral Angle

2）下反角

It is the angle when the tips of the wing are lower than the roots.

当指尖端低于根部所成的角为下反角。

Dihedral angle
上反角

Anhedral angle
下反角

Fig. 3-21   Dihedral and anhedral angle
图 3-21   上反角和下反角

3) Incidence Angle

3）机翼安装角

It is the angle between the wing chord and the fuselage centerline.

机翼安装角指机翼弦与机身中心线之间的角度。

3. Influence of Installation Parameters on Aircraft Performance

3. 安装参数对飞机性能的影响

Dihedral is used to promote the longitudinal and vertical stability of the aircraft.

Anhedral is used to promote the maneuverability by demoting the stability of the aircraft.

The incidence angle and the dihedral angle are important parameters that affect the flight performance of the aircraft. For modern civil transport aircraft, both parameters are determined in aircraft designing and manufacturing stages, and can not be adjusted after that stages. To ensure the airworthiness of the aircraft, both parameters should meet the requirements during the use and maintenance of the aircraft.

Wing Installation
Parameters (1)

上反角用来提高飞机的横侧向稳定性。

下反角用来降低飞机的稳定性，提高飞机的操纵性。

机翼的上反角和上反角是影响飞机飞行性能的重要参数。现代民用运输飞机的这两个参数是在飞机设计和制造阶段确定的，在飞机投入使用后无法调整。为确保飞机的适航性，这两个参数应满足飞机使用和维护期间的要求。

Wing Installation
Parameters (2)

New Words

biplane	['baɪpleɪn]	n.	双翼飞机
stack	[stæk]	n.	一叠，一摞
		v.	堆叠
monoplane	['mɒnəpleɪn]	n.	单翼飞机
configuration	[kənˌfɪɡjəˈreɪʃn]	n.	结构，构造，布局
mount	[maʊnt]	n.	托架
		v.	安装
low wing	[ləʊ wɪŋ]		下单翼
mid wing	[mɪd wɪŋ]		中单翼
approximately	[əˈprɒksɪmətli]	adv.	大概，大约
high wing	[haɪ wɪŋ]		上单翼
dihedral	[daɪˈhidrəl]	n.	上反角
		adj.	有两个平面的
anhedral	['ænhdrəl]	n.	下反角
		adj.	下反角的
lateral	['lætərəl]	n.	侧面
		adj.	侧面的，横向的
stability	[stəˈbɪləti]	n.	稳定性，稳定
incidence	['ɪnsɪdəns]	n.	发生范围
promote	[prəˈməʊt]	vt.	提高
maneuverability	[məˌnuːvərəˈbɪlɪti]	n.	操纵性
demote	[ˌdiːˈməʊt]	vt.	使降低
performance	[pərˈfɔːrməns]	n.	表现，性能
design	[dɪˈzaɪn]	n.	设计，布局
		vt.	设计
manufacturing	[ˌmænjuˈfæktʃərɪŋ]	n.	制造业
		v.	生产
adjust	[əˈdʒʌst]	v.	调整，调节
ensure	[ɪnˈʃʊr]	v.	确保，保证
airworthiness	['erwɜːrðinəs]	n.	适航性
requirement	[rɪˈkwaɪərmənt]	n.	要求，必要条件
maintenance	['meɪntənəns]	n.	维修，维护，保养

Q&A

The following questions are for you to answer to assess the learning outcomes.

(1) Please describe the difference between biplane and monoplane.

(2) How to distinguish the high wing, the mid wing and the low wing.

(3) What is the effect of the upper anti-angle?

空气动力
Aerodynamic Force

Contents

1) Lift

2) Drag

3) Aerodynamic force factors

4) Center of pressure(CP) and Aerodynamic center(AC)

学习内容

1）飞行升力

2）飞行阻力

3）空气动力的影响因素

4）压力中心和空气动力中心

任务 1 飞行动力
Task 1 Lift

 Contents

1) Aerodynamic force

2) Lift

3) Angle of attack

4) Stagnation point

Learning Outcomes

1) Master the forces of the aircraft in flight

2) Master the definition of angle of attack

3) Understand the principle of generated lift in flight

4) Master the composition of aerodynamic force

5) Analyze the changes of physical parameters in aerodynamics

6) Solve relevant problems in aerodynamics with lift

7) Cultivate professional qualities of rigor, carefulness, and ability to express, coordinate, and communicate effectively

任务内容

1）空气动力

2）升力

3）迎角

4）最低压力点

任务目标

1）掌握飞机飞行中的受力情况

2）掌握迎角的定义

3）理解飞行中升力产生的原理

4）掌握空气动力的组成

5）分析空气动力学物理参数的变化及原因

6）运用升力解决相关的空气动力学问题

7）培养严谨、细心的职业素养，以及有效表达、协调和沟通的能力

Learning Guide

Any aircraft must generate a lift greater than its own gravity in order to take off and fly, which is the basic principle of aircraft flight. Aircraft can be divided into two categories, lighter than aircraft and heavier than aircraft. Lighter than aircraft such as balloons, airships, etc., the main part of which is a large air bag filled with gases smaller than density of the air (such as hot air, hydrogen, etc.), such as our toy hydrogen balloons we played when we were young, we can rely on the static buoyancy of air to rise it into the air. More than a thousand years ago, our ancestors invented the Kongming lantern, a delicate instrument that uses hot air to soar, which can be considered the ancestor of lighter aircraft than air. However, for aircraft heavier than the air, such as airplanes, what force does it rely on to fly into the sky?

 课文

Aerodynamic Force
空气动力

The aircraft is flying in the air, and there is a force called lift that counteracts the gravity and keeps the aircraft in the air. To let an aircraft fly forward in the air, a power device is used to generate the thrust and counteract the drag (Fig. 4-1).

The force of the air acting on an aircraft when flying is called aerodynamic force, referred as R.

The aerodynamic forces on the aircraft totally act on the center of pressure. The aerodynamic force is composed of two components, lift and drag. The lift (L) is perpendicular to the relative airflow, and counteracts the force of gravity and keeps the aircraft flying. The drag (D) is parallel to the relative airflow, and counteracts the thrust to decelerate the aircraft.

飞机在空中飞行，有一种称为升力的力可以抵消重力，使飞机保持在空中。为了让飞机在空中向前飞行，动力装置被用来产生推力和抵消阻力（如图 4-1 所示）。

作用在飞机上的空气力称为空气动力，通常用 R 表示。

飞机上的空气动力完全作用在压力中心。空气动力可以分解为两个分量：升力和阻力。升力（L）垂直于相对气流，用于抵抗重力并保持飞机飞行。阻力（D）平行于相对气流，用于抵消推力以使飞机减速。

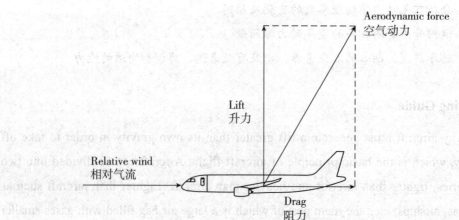

Fig. 4-1　Aerodynamic force on an aircraft
图 4-1　飞机上的空气动力

1. Lift

1. 升力

The aerodynamic forces acting on the upper and lower surfaces of the wing formed the general aerodynamic force, whose direction is mainly upward and slightly backward.

The component of this force in the direction perpendicular to the relative airflow is lift, and the component parallel to the relative flow is drag. Because the direction of the aerodynamic force acting on the wing is generally vertical, the lift is much greater than the drag.

The lift of an aircraft is mainly generated by the wings. When the air flows through the airfoil surface, the pressure difference between the upper and lower surfaces of the wing produces the lift.

作用在机翼上下两个表面的气动力形成了机翼气动力的合力，它主要是向上并稍向后倾的力。

该力在垂直于相对气流方向上的分量为升力，在平行于相对气流方向的分量为阻力。作用在机翼上的气动力方向基本上是垂直的，因此机翼产生的升力远大于阻力。

飞机的升力主要由机翼产生。当空气流经机翼表面时，机翼上下表面之间的压力差产生升力。

2. Angle of Attack

2. 迎角

The angle between the chord line and the relative airflow is the angle of attack (AoA), referred as α (Fig. 4-2). When the relative airflow blows from below of the wing chord, the AoA is positive, otherwise negative.

弦线和相对气流之间的角度是迎角（AoA），通常用 α 表示（图 4-2）。当相对空气从翼弦下方吹来时，迎角为正值，否则为负值。

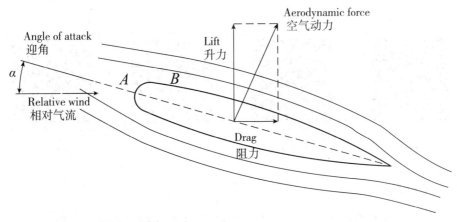

Fig. 4-2　Angle of attack
图 4-2　机翼的迎角

When the air flows through the airfoil at positive angle of attack, the flow tube on the upper surface of the wing becomes thinner and the streamline becomes denser. At the same time, the flow tube on the lower surface of the airfoil becomes wider and the streamline becomes sparser. The compressibility and viscosity can be ignored in the low speed air. According to Bernoulli principle, since the total pressure is a constant, the air speed on the upper surface of the airfoil increases and its static pressure decreases, the air speed on the lower surface of the airfoil decreases and its static pressure increases.

当空气以正迎角流过机翼时，机翼上表面的流管变得更细，流线变得更密集。同时，机翼下表面的流管变粗，流线稀疏。在低速空气中，空气的压缩性和黏度可以忽略。根据伯努利原理，由于总压力是恒定的，机翼上表面的空气速度增加，其静态压力降低，机翼下表面上的速度降低，其静态压力升高。

3. Stagnation Point

3. 最低压力点

The pressure on the upper surface of the airfoil is negative, and the pressure on the lower surface of the airfoil is positive. Pressure distribution on the airfoil surface is shown in Fig. 4–2. At the leading edge of the airfoil, point A shown in Fig. 4–2, the air speed decreases to zero and the static pressure reaches the maximum. Point A is called the stagnation point. On the upper surface of the airfoil, point B shown in Fig. 4–2, where the air speed reaches the maximum, gets the maximum dynamic pressure and the minimum static pressure. Point B is the lowest pressure point.

机翼上表面的压力为负，机翼下表面的压力为正。机翼表面的压力分布如图 4-2 所示。在机翼前缘 A 点处，空气速度降至零，正压达到最大值，该点为驻点。在机翼上表面 B 点处，空气速度达到最大值时，获得最大动态压力，此时静态压力最小，此点称为最低压力点。

Lift (1)　　　　　Lift (2)　　　　　Lift (3)　　　　　Lift (4)

 New Words

lift	[lɪft]	*n.*	（飞行时的）增升力，升力
		v.	举起
generate	['dʒenəreɪt]	*vt.*	生成，产生

thrust	[θrʌst]	*n.* 推力，驱动力
		v. 推动
aerodynamic	[ˌerəʊdaɪ'næmɪk]	*adj.* 空气动力学的，流线型的
resolve	[rɪ'zɒlv]	*n.* 决定
		v. 解决
center of pressure	['sentər əv 'preʃər]	压力中心
component	[kəm'pəʊnənt]	*n.* 成分，部件
		adj. 成分的，组成的
drag	[dræg]	*n.* 空气阻力
		v. 拖曳，拖
vertical	['vɜːrtɪkl]	*n.* 垂直线，垂直位置
		adj. 垂直的，纵向的
parallel	['pærəlel]	*n.* 平行
		adj. 平行的，并行的
		vt. 与……相似
		adv. 平行地，并列地
decelerate	[ˌdiː'seləreɪt]	*v.* 减速，变慢
angle of attack	['æŋgl əv ə'tæk]	迎角，仰角
blow	[bləʊ]	*n.* 吹
		v. 吹，刮
positive	['pɒzətɪv]	*n.* 正的，阳极
		adj. 阳性的，正数的，正电的
negative	['negətɪv]	*n.* 负的，阴极
		adj. 负的，负极的
		vt. 否定，拒绝
difference	['dɪfrəns]	*n.* 差别，差异，差动
		vt. 辨别，区分
viscosity	[vɪ'skɒsəti]	*n.* 黏性，黏度
stagnation	[stæg'neɪʃ(ə)n]	*n.* 失速，停滞

 Q&A

The following questions are for you to answer to assess the learning outcomes.

(1) When is the angle of attack positive and negative?

(2) Describe the principle of lift generation.

(3) Describe the pressure distribution on the wing surface.

 Extended Reading

What Is Lift?

Lift is the force that directly opposes the gravity of an airplane and holds the airplane in the air. Lift is generated by every part of the airplane, but most of the lift on a normal airliner is generated by the wings. Lift is a mechanical aerodynamic force produced by the motion of the airplane through the air. Because lift is a force, it is a vector quantity, having both a magnitude and a direction associated with it. Lift acts through the center of pressure of the object and is directed perpendicular to the flow direction. There are several factors which affect the magnitude of lift.

1. How Is Lift Generated?

There are many explanations for the generation of lift found in encyclopedias, in basic physics textbooks, and on websites. Unfortunately, many of the explanations are misleading and incorrect. Theories on the generation of lift have become a source of great controversy and a topic for heated arguments.

Lift occurs when a moving flow of gas is turned by a solid object. The flow is turned in one direction, and the lift is generated in the opposite direction, according to Newton's third law of action and reaction. Because air is a gas and the molecules are free to move about, any solid surface can deflect a flow. For an aircraft wing, both the upper and lower surfaces contribute to the flow turning. Neglecting the upper surface's part in turning the flow leads to an incorrect theory of lift.

2. No Fluid, No Lift

Lift is a mechanical force. It is generated by the interaction and contact of a solid body with a fluid (liquid or gas). It is not generated by a force field, in the sense of a gravitational field, or an electromagnetic field, where one object can affect another object without being in physical contact. For lift to be generated, the solid body must be in contact with the fluid: No fluid, no lift. The space shuttle does not stay in space because of lift from its wings but because of orbital mechanics related to its speed. Space is nearly a vacuum. Without air, there is no lift generated by the wings.

3. No Motion, No Lift

Lift is generated by the difference in velocity between the solid object and the fluid. There must be motion between the object and the fluid: No motion, no lift. It makes no difference whether the object moves through a static fluid, or the fluid moves past a static solid object. Lift acts perpendicular to the motion. Drag acts in the direction opposed to the motion.

任务 2 飞行阻力
Task 2 Drag

 Contents

1) Boundary layer

2) Transition of boundary layer

3) Boundary layer separation

4) Friction drag

5) Pressure drag

6) Interference drag

7) Wingtip vortex

8) Downwash

9) Induced drag

10) Drags when flying at low speeds

11) Drags vs. Flight speeds

Learning Outcomes

1) Master the definition of boundary layer and understand its impact on flight

2) Master the causes and influencing factors of each drag

3) Understand the measures to reduce the drags

4) Master the characteristics of each drag and total drags

5) Cultivate professional qualities of rigor, carefulness, and ability to express, coordinate, and communicate effectively

 任务内容

1) 附面层

2) 附面层的转捩

3）附面层分离

4）摩擦阻力

5）压差阻力

6）干扰阻力

7）翼尖涡

8）气流下洗

9）诱导阻力

10）低速飞行时的阻力

11）阻力与飞行速度的关系

 任务目标

1）掌握附面层的定义，理解其对飞行的影响

2）掌握每种阻力的产生原因和影响因素

3）理解减小每种阻力的措施

4）掌握每种阻力和总阻力的特点

5）培养严谨、细心的职业素养，以及有效表达、协调和沟通的能力

Learning Guide

We know that the forces acting on the aircraft during flight are balanced. The gravity of an aircraft is balanced with the lift generated by the aircraft, and the function of the aircraft's engine is to overcome the drag of the aircraft and push it forward, causing the aircraft to move relative to the air, thereby generating lift. Everyone must think, the power of an aircraft engine is so high, is there so much drag on the aircraft? Indeed, while flying at high speeds, airplanes will experience significant resistance due to different reasons.

 课文 1

1. Boundary Layer
1. 附面层

Air is viscous and clings to the surface over which it flows. At the surface, the air particles are slowed to near–zero relative velocity. Above the surface, the retarding forces lessen progressively, until slightly above the surface the particles have the full velocity of the airstream. The thin viscous region adjacent to the body is called the boundary layer.

According to the flow state in the boundary layer, it can be divided into both laminar boundary layer and turbulent boundary layer.

In the boundary layer, the air friction will decrease the air speed, and the decreased dynamic energy cannot be transformed into pressure energy. It has no impact on the pressure on the surface.

The flow in the boundary layer has no influence on the one in the outer layer. However, the outer layer plays a defining role in the boundary layer (Fig. 4-3).

空气具有黏性。当空气流经机体表面时，靠近体表的空气将被机体表面拖动且相对速度降至零，进而拖动邻近的外部空气以降低其速度，直至空气速度从零逐渐增加到外部气流速度。流速由零增加到外部流速的这一层紧贴机体表面薄薄的空气层被称为附面层，或者边界层。

根据附面层中空气的流动状态，可将附面层分为层流附面层和湍流附面层。

在附面层中，流体流速的降低是由空气摩擦引起的，而降低的动力能不会转化为压力能，所以机体表面处各点气流的压强并不会因为流速的减小而增加。

附面层对其外部流体流动的影响可以忽略，但外部流体对附面层内的流体流动起决定作用（图 4-3）。

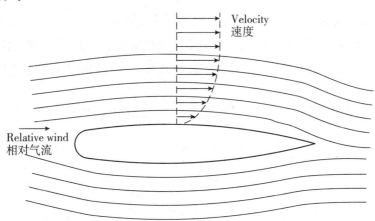

Fig. 4-3　The flow in the boundary layer
图 4-3　附面层内空气的流动

When the air flows through the airfoil surface, the front section is a laminar boundary layer, in which the streamlines are smooth and regular, and the streamlines flow backward along the surface layer by layer. The air between the different layers are not mixed with each other. In the laminar boundary layer, the drag force between the flow layers is caused by viscosity.

In the rear section of the boundary layer, the air moves forward and irregularly. It is difficult to distinguish the flow layers. It becomes a turbulent boundary layer.

当空气流经翼型表面时，前部为层流附面层，其中流线平滑、规则，空气沿表面明显一层一层地向后流动。不同层之间的空气不相互混合。在层流附面层中，流动层之间的阻力由黏度引起。

在附面层的后部，空气向前流动，同时也上下乱窜，很难区分流动的层次。它被称为湍流附面层。

2. Transition of Boundary Layer

2. 附面层的转捩

The transition of boundary layer (Fig. 4-4) is the flow state change from laminar to turbulence. The transition regions are called transition sections.

The main reason why the transition from laminar state to turbulent state is that the longer the distance of air flows through, the thicker the boundary layer becomes, and the more unstable flow in the boundary layer. The body surface is rough and not flat (oil, dirt, protruding rivet head, skin joint), which disturbs the boundary layer, and finally leads to the transition of the boundary layer. Besides the thickness of the turbulence is thicker than that of the laminar, the drag effect of the turbulence is also much larger than that of the laminar.

附面层的转捩（图 4-4）是流动状态从层流到湍流的转变。流动状态的过渡区域称为转捩区。

气流通过的距离越长，附面层越厚，附面层中的流动越不稳定，机体表面的粗糙度（如油、灰尘、凸出的铆钉头、蒙皮接头）不断干扰附面层，最终导致附面层从层流状态过渡为湍流状态。除湍流的厚度比层流的厚度厚之外，湍流的阻力效应也比层流大得多。

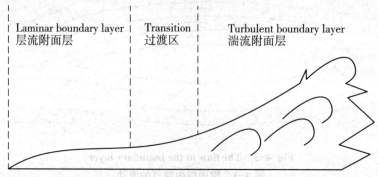

Laminar boundary layer
层流附面层　　Transition 过渡区　　Turbulent boundary layer 湍流附面层

Fig. 4-4　Transition of the boundary layer

图 4-4　附面层的转捩

3. Boundary Layer Separation

3. 附面层分离

From stagnation point to the lowest pressure point, the speed of the air movement gradually increases, and the static pressure gradually decreases. However, after the lowest pressure point, the speed of the air movement gradually decreases and the static pressure gradually increases, leading to an inverse pressure gradient region in which the rear pressure is greater than the front pressure. This is not good for the airflow because the viscous drag and the inverse pressure hinder the airflow in the boundary layer. Therefore, after entering the inverse pressure gradient region, the speed of the air movement decreases rapidly, and the bottom airflow will flow back under the inverse pressure, leading to the vortex. This phenomenon is called boundary layer separation (Fig. 4–5).

The vortices generated by the boundary layer separation are continuously blown away by the main airflow, and many new vortices are continuously generated from the surface, leading to wake.

从驻点到最低压力点，气流逐渐加速，静态压力逐渐降低。然而，从最低压力点向后，气流逐渐减速，静态压力逐渐增加，形成逆压梯度区，其中后压力大于前压力。因为黏性阻力和逆压阻力阻碍附面层内气流向后流动，这对附面层内气流的流动极为不利。因此，进入逆压梯度区后，流体流速迅速下降，底部气流将在逆压下产生回流，与顺流而下的气流相撞，从而形成旋涡。这种现象称为附面层分离（图4-5）。

主气流不断地吹走由附面层分离产生的旋涡，并从机体表面不断产生新的旋涡，从而形成尾迹。

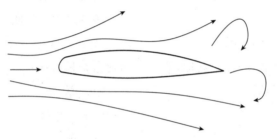

Fig. 4–5 Boundary layer separation
图4-5 附面层分离

Drag (1)   Drag (2)   Drag (3)

 New Words

adjacent	[əˈdʒeɪsnt]	adj.	相邻的，邻近的
boundary	[ˈbaʊndri]	n.	边界，界限
layer	[leɪər]	n.	层，表层，层次
		vt.	把……分层堆放
friction	[ˈfrɪkʃn]	n.	摩擦，摩擦力

ignore	[ɪgˈnɔːr]	vt.	忽视，对……不予理会
role	[rəʊl]	n.	作用，影响程度
laminar	[ˈlæmɪnər]	adj.	层状的；薄片状的
turbulent	[ˈtɜːrbjələnt]	adj.	混乱的，湍动的
backward	[ˈbækwərd]	adj.	向后的，朝后的
		adv.	向后，倒退地
distinctly	[dɪˈstɪŋktli]	adv.	显然，明显地
mix	[mɪks]	n.	混合，混杂
		v.	混合，(使)掺和
turbulence	[ˈtɜːrbjələns]	n.	湍流，紊流
rear	[rɪr]	n.	后部
		adj.	后面的，后部的
distinguish	[dɪˈstɪŋgwɪʃ]	v.	区分，辨别，分清
transition	[trænˈzɪʃn]	n.	过渡，转变
		v.	转变过程，过渡
region	[ˈriːdʒən]	n.	区域，部位
transition section	[trænˈzɪʃn ˈsekʃn]		过渡段，过渡区
thick	[θɪk]	n.	粗，厚
		adj.	厚的，粗的
		adv.	厚厚地
rough	[rʌf]	n.	草稿，草图
		v.	使粗糙，草拟
		adj.	粗糙的，不平滑的
		adv.	粗鲁地，粗野地
oil	[ɔɪl]	n.	油，润滑油
		vt.	给……加润滑油
dirt	[dɜːrt]	n.	污垢，污物
protrude	[prəʊˈtruːd]	vi.	突出，伸出，鼓出
rivet head	[ˈrɪvɪt hed]		铆钉头
skin	[skɪn]	n.	皮，薄皮
		v.	剥皮，擦破
disturb	[dɪˈstɜːrb]	vt.	打扰，干扰，妨碍，搅乱
gradually	[ˈgrædʒuəli]	adv.	逐步地，逐渐地，渐进地
inverse	[ˌɪnˈvɜːrs]	n.	反面，相反的事物

		vt.	使倒转，使颠倒
		adj.	相反的，反向的
gradient	['greɪdiənt]	*n.*	坡度，斜率，倾斜度
		adj.	倾斜的，步行的
rapidly	['ræpɪdli]	*adv.*	迅速地

boundary layer separation ['baʊndri 'leɪə(r) ˌsepə'reɪʃn] 附面层分离

wake	[weɪk]	*n.*	尾流，航迹
		v.	醒，醒来，唤醒

 ## Q&A

The following questions are for you to answer to assess the learning outcomes.

(1) Briefly describe the causes of the boundary layer.

(2) Briefly describe the characteristics of the laminar boundary layer.

(3) Briefly describe the characteristics of turbulent boundary layer.

(4) Why does the boundary layer change from laminar flow to turbulent flow?

(5) Briefly describe the boundary layer separation.

 ## 课文 2

1. Definition of Friction Drag
1. 摩擦阻力的定义

Friction drag is the drag caused by the viscosity of the air in the boundary layer. The aircraft surface gives the air a forward retarding force which will reduce the air's speed. Meanwhile, the air gives the aircraft a reverse force in the same direction. The reverse force is the friction drag.

摩擦阻力是由附面层中的空气黏度引起的阻力。飞机表面给空气向前的阻滞力，使其速度下降，同时空气在同一方向上给飞机一个反向力。这个反向力就是摩擦阻力。

2. Properties of Friction Drag
2. 摩擦阻力的性质

The drag of the turbulent boundary layer is greater than that of the laminar boundary layer. Therefore, the turbulent boundary layer will produce most of the friction drag (Fig. 4-6).

湍流附面层的阻力大于层流附面层的阻力。因此，与层流附面层相比，湍流附面层将产生大部分摩擦阻力（图 4-6）。

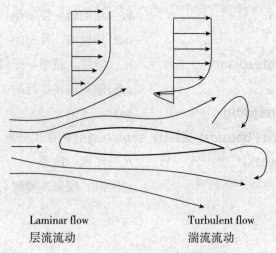

Laminar flow
层流流动

Turbulent flow
湍流流动

Fig. 4-6   Friction drag on the airfoil
图 4-6   机翼上的摩擦阻力

Besides the influence of the flow state in the boundary layer, the friction drag is influenced by the interface area between the aircraft and the airflow. The larger the interface area is, the greater the friction drag is.

Since most of the friction drag is generated in the turbulent boundary layer, the boundary layer is managed to keep in the laminar state as much as possible to reduce the friction drag.

Laminar airfoil is an effective airfoil to keep the boundary layer in laminar state, delaying the transition and separation of the boundary layer (Fig. 4-7).

除附面层的流动状态外，摩擦阻力还受物体与气流之间的界面面积的影响。界面面积越大，摩擦阻力越大。

由于大部分摩擦阻力是在湍流附面层中产生的，因此附面层应尽可能保持在层流状态，以减少摩擦阻力。

层流翼型是保持附面层处于层流状态、延迟附面层过渡和分离的有效翼型（图 4-7）。

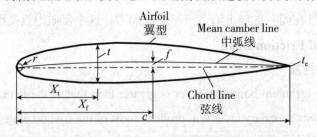

Fig. 4-7   Laminar airfoil
图 4-7   层流翼型

Some aerodynamic devices are installed on the wing surface to add dynamic energy into the boundary layer, which can also delay the boundary layer separation.

Keeping the aircraft surface smooth and clean is also an effective way to reduce friction drag.

In aircraft maintenance, the aircraft surface must be kept smooth and tidy, especially the main aerodynamic surfaces, such as the leading edge and upper surface of the wing and tail. Ensure that there is no dirt, scratch, dent or protrusion on the surface, pay attention to the quality of countersunk rivets and the smooth sealing of skin overlap joints. The interface area between the airframe and the airflow shall be minimized. When repairing and refitting the aircraft, the exposed area shall be minimized, otherwise the drag will increase and the aircraft will not meet the requirements of flight performance.

一些安装在机翼表面的气动装置能向附面层中添加动能，因而也可以延迟附面层分离。

保持飞机表面光滑、清洁也是减少摩擦阻力的有效方法。

在飞机维修中，飞机表面必须保持光滑和整洁，尤其是主要的气动表面，如机翼和尾翼的前缘和上表面。确保表面无污垢、划痕、凹痕或凸起，注意沉头铆钉的质量和蒙皮搭接处的平滑密封。机身和气流之间的接触面积应最小化。飞机在被修理和改装时，应尽量减少暴露面积，否则阻力会增加，将无法满足飞行性能要求。

Drag (4)

Drag (5)

Drag (6)

 New Words

friction	[ˈfrɪkʃn]	n.	摩擦；摩擦力
bottom	[ˈbɒtəm]	n.	底部，最下部
		v.	给……装上底，降到最低
		adj.	底部的，最后的
interface	[ˈɪntərfeɪs]	n.	界面，接口，接合点
		v.	接合，连接
generate	[ˈdʒenəreɪt]	vt.	生成，产生，引起
effective	[ɪˈfektɪv]	n.	有效
		adj.	有效的，实际的
delay	[dɪˈleɪ]	n.	延误，延期，延迟
		v.	延迟，推迟，延期

device	[dɪ'vaɪs]	n.	装置，仪器，器具
install	[ɪn'stɔːl]	vt.	安装，设置
smooth	[smuːð]	n.	平整，平坦
		adj.	平整的，平坦的，平滑的，光滑的
		vt.	使平整，使平坦
		adv.	平稳地，顺利地
maintenance	['meɪntənəns]	n.	维修，维护，保养，维持
tidy	['taɪdi]	v.	整理，使整洁，使有条理
aerodynamic surface	[ˌerəʊdaɪ'næmɪk 'sɜːrfɪs]		空气动力面
tail	[teɪl]	n.	尾，尾巴，机尾
ensure	[ɪn'ʃʊr]	v.	确保，保证，担保
dirt	[dɜːrt]	n.	污垢，污物，
scratch	[skrætʃ]	n.	划痕，划伤
		v.	擦破，刮坏
dent	[dent]	n.	凹痕，凹坑，凹部
		vt.	使凹陷，使产生凹痕
protrusion	[prəʊ'truːʒn]	n.	突出物，凸起，伸出
sealing	['siːlɪŋ]	n.	密封装置
		v.	密封，封上
quality	['kwɒləti]	n.	质量，品质
		adj.	优质的，高质量的
countersunk	['kaʊntərsʌŋk]	v.	钻孔装埋，打埋头孔
		adj.	沉头的
overlap	[ˌəʊvər'læp]	n.	重叠部分
		v.	重叠，部分重叠，交叠
airframe	['ɛrˌfreɪm]	n.	机身
repairing	[rɪ'perɪŋ]	v.	修理（repair 的现在分词），修补
refitting	[ˌriː'fɪtɪŋ]	v.	整修，改装（refit 的现在分词）
expose	[ɪk'spəʊz]	n.	暴露
		vt.	露出，显露
flight performance	[flaɪt pər'fɔːrməns]		飞行性能

 Q&A

The following questions are for you to answer to assess the learning outcomes.

(1) What is friction drag?

(2) What factors will affect the friction drag?

(3) What aspects should be paid attention to in aircraft maintenance to keep the smoothness of aircraft surface?

 课文 3

1. Definition of Pressure Drag

1. 压差阻力的定义

When the air flows through the airfoil surface, the air speed drops to zero at the stagnation point on the wing leading edge, this is the maximum pressure point. After the maximum pressure point, the boundary layer is separated by the action of the inverse pressure region, and a low pressure wake is formed at the trailing edge of the wing. In this way, the pressure in the leading edge area of the airfoil is greater than that in the trailing edge area, and the pressure difference between the front and rear forms pressure drag (Fig. 4-8).

当空气流经机翼表面时，机翼前缘驻点处的空气速度降至零，即最大压力点。在最大压力点之后，由于逆压区域的作用，附面层分离，机翼后缘形成低压尾流。这样，机翼前缘区域的压力大于后缘区域的压力，前后之间的压力差形成压差阻力（图 4-8）。

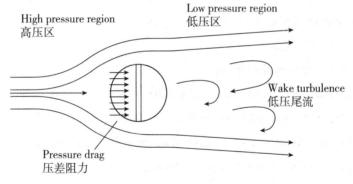

Fig. 4-8　Generation of pressure drag

图 4-8　压差阻力的产生

2. Properties of Pressure Drag

2. 压差阻力的性质

The pressure drag is effected by the frontal area, the shape of the airfoil, and the angle of attack (Fig. 4-9).

The bigger the frontal area, the larger the pressure drag. The smaller the angle of attack, the lower the pressure drag.

The more streamlined the shape of the airfoil, the lower the pressure drag.

压差阻力受迎风面积、机翼形状和机翼迎角的影响（图4-9）。

迎风面积越大，压差阻力越大；迎角越小，压差阻力越小。

机翼形状越流线型，压差阻力越低。

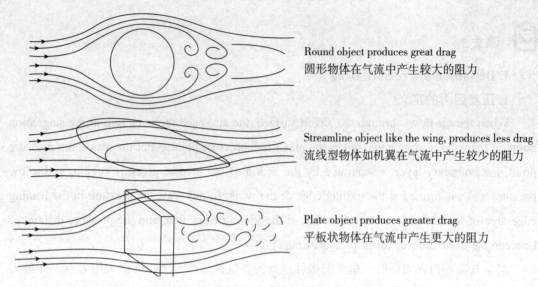

Round object produces great drag
圆形物体在气流中产生较大的阻力

Streamline object like the wing, produces less drag
流线型物体如机翼在气流中产生较少的阻力

Plate object produces greater drag
平板状物体在气流中产生更大的阻力

Fig. 4-9　Factors of pressure drag
图4-9　压差阻力的影响因素

3. The Methods to Reduce the Pressure Drag
3. 减少压力阻力的方法

Reduce the frontal area of the aircraft.

The cross sectional shape of the fuselage shall be circular or approximately circular.

The shape of all parts exposed in the airflow shall be streamlined.

During flight, the axis of aircraft parts shall be parallel to the airflow direction. A certain installation angle (Fig. 4-10) is adopted for the wing of civil transport aircraft to make the fuselage axis parallel to the incoming flow and to reduce the pressure drag while the wing generates the required lift during cruise flight.

减小飞机的前部面积。

飞机机身的横截面形状应为圆形或近似圆形。

暴露在气流中的飞机所有部件的形状应为流线型。

飞行期间，飞机部件的轴线应平行于气流方向。民用运输机的机翼采用一定的安装角（图4-10），使机身轴线与来流平行，以减少机翼在巡航飞行期间产生所需升力时的压差阻力。

Drag (7)

Drag (8)

Fig. 4-10　Wing installation angle
图 4-10　机翼安装角

 New Words

pressure drag	['preʃər dræg]		压差阻力
differential	[ˌdɪfə'renʃl]	*n.*	差别，微分
		adj.	特异的，差别的
frontal	['frʌntl]	*n.*	正面
		adj.	正面的，前部的
relative	['relətɪv]	*n.*	同类事物
		adj.	相对的，比较的
streamline	['striːmlaɪn]	*n.*	流线型，流线
		vt.	使……成流线型
		adj.	流线型的
circular	['sɜːrkjələr]	*adj.*	圆形的，环形的
civil transport aircraft	['sɪvl 'trænspɔːt 'eəkrɒft]		民用运输机
generate	['dʒenəreɪt]	*v.*	生成，产生，引起
cruise	[kruːz]	*n.*	巡游，乘船游览
		v.	巡航

 Q&A

The following questions are for you to answer to assess the learning outcomes.

(1) Describe the definition of pressure drag.

(2) Where is the minimum and maximum pressure on the airfoil surface?

(3) What factors will affect the pressure drag?

(4) Why should the wing shape be streamlined?

(5) Briefly describe the role of wing installation angle.

课文 4

1. Interference Drag
1. 干扰阻力

The drag of the whole aircraft is not the sum of the separate drag forces generated by each component, there is an extra drag, which is the interference drag. Interference drag is caused by the mutual interference of the air flow through the components on the aircraft at joint regions (Fig. 4-11).

The interference drag is effected by the mutual alignment of the components, and the shape of the flow tube formed at the joint of the components.

整个飞机的阻力不是每个部件产生的阻力之和，还有一个额外的阻力，即空气流经每个部件的接合区域时，气流相互干扰产生的阻力，称为干扰阻力（图 4-11）。

干扰阻力受部件的位置关系和部件连接处所形成流管的形状影响。

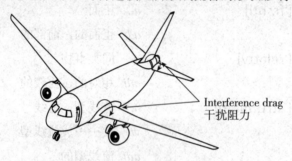

Fig. 4-11 Interference drag at the wing fuselage joint
图 4-11 机翼机身连接处的干扰阻力

2. Methods to Reduce the Interference Drag
2. 减少干扰阻力的方法

Arrange the positions configuration of the components properly, e.g. the interference drag of the mid wing aircraft is the smallest, the low wing aircraft is the largest, the high wing is moderate.

Install fairing at the joint of components to make the joint smooth to reduce the interference drag.

合理安排部件的位置，如中翼的干扰阻力最小，下单翼最大，高翼居中。

在部件的上单翼安装整流罩，使接头更加平滑，以减少干扰阻力。

Drag (9)

New Words

mutual	['mjuːtʃuəl]	*adj.*	相互的，彼此的
interference	[ˌɪntər'fɪrəns]	*n.*	干扰，干涉
sum	[sʌm]	*n.*	总和，和
		v.	归纳，总计
component	[kəm'pəʊnənt]	*n.*	组成部分，成分，部件
		adj.	成分的，组成的
extra	['ekstrə]	*adj.*	附加的，额外的
arrange	[ə'reɪndʒ]	*v.*	排列，布置
properly	['prɒpərli]	*adv.*	正确地，适当地
configuration	[kənˌfɪgjə'reɪʃn]	*n.*	配置，结构，构造，布局
fairing	['ferɪŋ]	*n.*	整流罩，整流装置

 Q&A

The following questions are for you to answer to assess the learning outcomes.

(1) Describe the definition of the interference drag.

(2) Briefly describe how the interference drag is generated.

(3) How to reduce the interference drag?

 课文 5

1. Wingtip Vortex

1. 翼尖涡

The lift is generated by the pressure difference between the upper and lower wing surfaces. This pressure difference will make the airflow an extra lateral movement from the high pressure area on the lower wing surface to the low pressure area on the upper wing surface through the tip of the wing, as shown in Fig. 4–12.

Fig. 4–12 Pressure difference between upper and lower surface of the wing

图 4–12 机翼上下翼面的压力差

In this way, the airflow on the lower surface of the wing is deflected from the wing root to the wingtip, and the airflow on the upper surface of the wing is deflected from the wingtip to the wing root, as shown in Fig. 4–13.

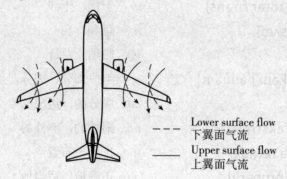

- - - Lower surface flow
下翼面气流

—— Upper surface flow
上翼面气流

Fig. 4–13 Airflow direction of the wing upper and lower surface
图 4–13 机翼上下翼面的气流方向

The wingtip vortex which is rotating from bottom to top is generated in the tip of the wing, as shown in Fig. 4–14.

升力产生的原因是上下翼面之间的压力差。该压力差将使空气在沿翼面向后流动的同时，从下翼面的高压区域侧向流向上翼面的低压区域，如图 4–12 所示。

这样，下翼面的气流从机翼根部偏转到机翼尖部，上翼面的气流从机翼尖部偏转到机翼根部，如图 4–13 所示。

在机翼尖部形成了由下而上旋转的翼尖涡，如图 4–14 所示。

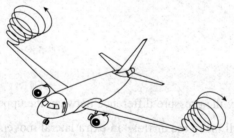

Fig. 4–14 Wingtip vortex
图 4–14 翼尖涡

2. Down Wash
2. 气流下洗

Due to the effect of wingtip vortex, the air flows backward and downward at the same time. This downward flow in the vertical direction is called down wash. Its speed is called down wash speed (Fig. 4–15).

受翼尖涡流的影响，机翼上下翼面的气流在向后流动的同时出现了向下流动的趋势。这种垂直方向的向下流动称为气流下洗。其速度称为向下洗速度（图 4–15）。

Fig. 4-15　Down wash of the wing

图 4-15　机翼气流下洗

3. Induced Drag

3. 诱导阻力

The airflow is inclined downward to the relative aivflow because of the down wash. The lift is also inclined backward at the same time. In this way, a component force of the lift plays the role of obstructing the flight forward, and the component force is called induced drag (Fig. 4-16).

由于下洗，气流相对于流入方向向下倾斜，同时其产生的升力会向后倾斜。这样，现在产生的升力在水平方向的分力起阻力作用。这种阻碍飞行的反向力被称为诱导阻力（图4-16）。

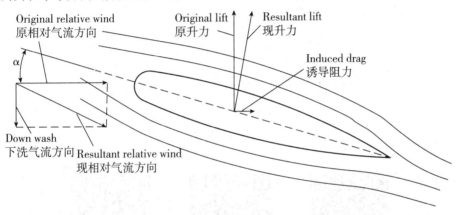

Fig. 4-16　Induced drag generation

图 4-16　诱导阻力的产生

4. Properties of Induced Drag

4. 诱导阻力的性质

Induced drag is a drag associated with the lift. The greater the pressure difference between the upper and lower surface of the wing, the greater the lift and the greater the induced drag.

诱导阻力是与升力相关的阻力。机翼上下表面之间的压差越大，升力越大，诱导阻力越大。

5. Methods to Reduce Induced Drag

5. 减少诱导阻力的方法

(1) The wing planform with small induced drag is adopted. For example, the wings with large aspect ratio will reduce the proportion of the wingtip are in the total wing area, leading to reducing the induced drag. The wings with large aspect ratio are often adopted in low speed

aircraft.

(2) The winglet at the tip of the wing is adopted to prevent the air's lateral movement from the lower surface to the upper surface of the wing, leading to reducing the induced drag (Fig. 4–17). Winglet plays an obvious role in reducing induced drag, saving fuel and increasing flight range.

（1）采用诱导阻力小的机翼形状，如增加机翼的展弦比，可以降低翼尖部位的面积在机翼总面积中所占的比例，从而减小诱导阻力。低速飞机大多采用大展弦比的机翼。

（2）采用翼尖小翼防止气流从机翼下表面横向流向上表面，从而减少诱导阻力（图 4–17）。小翼在减少诱导阻力、节省燃油和增加飞行范围方面起着明显的作用。

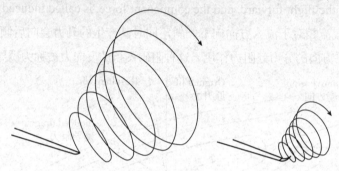

Fig. 4–17　Winglet

图 4–17　小翼

Drag (10)　　　　　　Drag (11)　　　　　　Drag (12)

 New Words

induced drag	[ɪnˈduːst dræg]		诱导阻力
associate	[əˈsəʊsieɪt]	*n.*	联合
		v.	联合
		adj.	副的，联合的
rotate	[ˈrəʊteɪt]	*v.*	（使）旋转，转动
down wash	[daʊn wɒʃ]		下洗，下冲气流
incline	[ɪnˈklaɪn]	*n.*	倾斜，倾侧
		v.	倾斜，（使）倾侧

hinder	['hɪndər]	*vt.* 阻碍，妨碍，阻挡
		adj. 后面的
winglet	[wɪŋ'lit]	*n.* 小翅，小翼
fuel	['fjuːəl]	*n.* 燃料，燃油
		v. 加油
flight range	[flaɪt reɪndʒ]	航程

 Q&A

The following questions are for you to answer to assess the learning outcomes.

(1) Describe the formation principle of the wingtip vortex.

(2) Briefly describe the direction of wingtip vortex.

(3) Describe the formation principle of down wash.

(4) Describe the formation principle of induced drag.

(5) How to reduce induced drag?

 课文 6

1. Drags When Flying at a Low Speed

1. 低速飞行时的阻力

When flying at a low speed, the aircraft drags are composed of friction drag, pressure drag, interference drag and induced drag.

Friction drag, pressure drag and interference drag are collectively referred to as parasite drag. The total drags of the aircraft is the sum of the parasite drag and the induced drag when flying.

The contribution of the four types of drags varies with the flight speed and the angle of attack.

低速飞行时，飞机阻力由摩擦阻力、压差阻力、干扰阻力和诱导阻力组成。

摩擦阻力、压差阻力和干扰阻力统称为废阻力。飞行期间的总阻力是诱导阻力和废阻力之和。

这四种阻力的作用随飞行速度和迎角的变化而变化。

2. Drags vs. Flight Speed

2. 阻力与飞行速度的关系

The relationship between drag and flight speed are as follows (Fig. 4–18).

(1) When flying at a low speed, the aircraft shall fly at a large angle of attack to get enough lift. The pressure difference between the upper and lower surfaces of the wing is large due to large

AoA, resulting in strong wingtip vortex and a large induced drag. When flying at a high speed, the aircraft flies at a small AoA, the pressure difference between the upper and lower surfaces of the wing decreases, resulting in weak tip vortex and a small induced drag. Therefore, the induced drag decreases gradually with the increase of the flight speed.

(2) The parasite drag is caused by the viscosity of the air. The higher the flight speed, the greater the drag generated by the viscosity of the air, and the greater the parasite drag. The parasite drag increases with the increase of the flight speed. When the flight speed is low, the induced drag is greater than the parasite drag, and the former plays a dominant role in the total drags. With the increase of flight speed, the induced drag gradually decreases, while the parasite drag increases, the parasite drag plays a dominant role.

(3) At the intersection point of the induced drag curve and the parasite drag curve, the total drag is the smallest, the flight speed at this point is called favorable flight speed.

阻力与飞行速度的关系如下（图 4-18）。

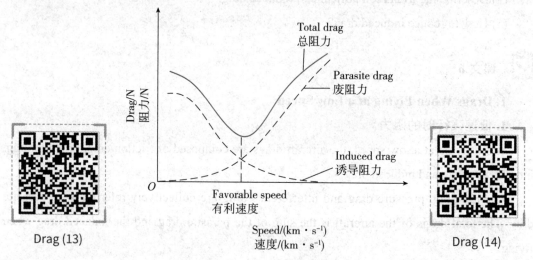

Fig. 4-18　Drags vs. Flight speed

图 4-18　阻力与飞行速度的关系

（1）低速飞行时，飞机应以大迎角飞行以获得足够的升力。由于大的迎角，机翼上下表面之间的压力差很大，从而产生强大的翼尖涡和较大的诱导阻力。当飞机高速飞行时，飞机以小迎角飞行，机翼上下表面之间的压差减小，从而产生微弱的翼尖涡和较小的诱导阻力。因此，诱导阻力随着飞行速度的增加而逐渐减小。

（2）废阻力是由空气黏度引起的阻力。飞行速度越高，气流产生的阻力越大，废阻力越大。废阻力随着速度的增加而增加。

（3）当飞行速度较低时，诱导阻力大于废阻力，并在总阻力中起主导作用。随着飞行速度的增加，诱导阻力逐渐减小，而废阻力增加，废阻力开始起主导作用。

（4）在诱导阻力曲线和废阻力曲线的交点处，总阻力最小，此时的飞行速度最佳，称为有利飞行速度。

 New Words

parasite drag	['pærəsaɪt dræg]	废阻力
vary	['veri]	v. 变化，不同
gradually	['grædʒuəli]	adv. 逐步地，逐渐地，渐进地
dominant	['dɒmɪnənt]	n. 主因，要素
		adj. 占主导地位，首要的
intersection	[ˌɪntər'sekʃn]	n. 交点，交汇点
curve	[kɜːrv]	n. 曲线，弧线
		v. 沿曲线运动，呈曲线形
favorable	['feɪvərəbəl]	adj. 有利的

 Q&A

The following questions are for you to answer to assess the learning outcomes.

(1) What are the drags composed of when flying?

(2) How do the aircraft drags change with the angle of attack?

(3) What is the cause of parasite drag?

(4) Under what circumstances will the total drag on an aircraft be minimal?

 Extended Reading

Aircraft Drag Reduction—a Review

Aerodynamic drag is historically and conveniently separated into pressure or form drag (including interference and roughness drag), drag due to lift, shock or compressibility drag and viscous or skin friction drag. Except for helicopters and military aircraft with external stores, which can still exhibit appreciable levels of pressure drag, cruise drag for most subsonic aircraft consists primarily of friction drag and drag due to lift. For supersonic cruise aircraft, shock drag can be the same order as (vortex) drag due to lift and friction drag.

The importance and possibilities for viscous drag reduction were seriously identified in the late 1930s, primarily as a result of two developments: Successful drag "clean-up" efforts which minimized pressure drag, thereby enhancing the importance of (residual) viscous drag, and the realization, via development of low disturbance facilities and flight transition measurements, that turbulent flow was not necessarily given beyond a Reynolds number of order 2×10^5. Such a low transition Reynolds number was common in the wind tunnels of the period, which typically exhibited stream turbulence levels on the order of 1% or greater. In fight and low disturbance tunnels, with stream disturbance levels on the order of 0.05%, transition could occur well beyond Reynolds numbers of order 2×10^6.

The earliest research in aeronautical viscous drag reduction addressed the issues of transition delay, initially via favourable pressure gradients on the essentially unswept wings of the day. Later, in the 1950s and 1960s, suction was utilized in research efforts to address the cross-flow instability problem endemic on swept wings. This early research on transition delay was termed "laminar flow control" (LFC), with "natural" laminar flow defined by pressure gradient controlled/delayed transition and "forced" or active laminar flow obtained via suction. This technology offered large gains in aircraft performance and was actively pursued, at various times, in many countries, e.g. the United States, Britain, France, Germany, Japan and Russia. This research demonstrated that, in carefully controlled experiments, transition could be delayed for appreciable distances with consequent large decreases in viscous drag (compared to the turbulent level). However, the critical (for application) maintenance and reliability issues were never, at least up to the mid 1960s, successfully addressed. Various "real world" problems, such as insect debris, other roughnesses and occurrence of waviness under loading, all exacerbated, initially, by the low cruise altitude/high unit Reynolds number prevalent in the 1940s and early 1950s (and later by wing sweep), kept LFC in the category of a "laboratory curiosity". The continued availability of inexpensive petroleum in the 1960s, coupled with these unresolved reliability and maintainability issues, caused an essential hiatus in LFC research from the mid 1960s to the mid 1970s.

任务 3　空气动力的影响因素
Task 3　Aerodynamic Force Factors

 Contents

4) Lift drag ratio vs. AoA

5) Stall

6) Stall speed

 Learning Outcomes

1) Master the expression of lift formula and drag formula

2) Master the trend and cause of lift curve, drag curve and lift drag ratio curve

3) Master the physical meaning expressed by the special points on the lift curve, drag curve and lift drag ratio curve

4) Master the phenomenon and cause of stall

5) Solve the aerodynamic problems by the lift/drag formula and the corresponding curve

6) Cultivate professional qualities of rigor, carefulness, and ability to express, coordinate, and communicate effectively

任务内容

1）升力 / 阻力公式

2）升力 / 阻力系数

3）升力 / 阻力系数与迎角的关系

4）升阻比与迎角的关系

5）失速

6）失速速度

任务目标

1）掌握升力公式和阻力公式的表达

2）掌握升力曲线、阻力曲线、升阻比曲线的趋势和变化原因

3）掌握升力曲线、阻力曲线、升阻比曲线上的特殊点所表达的物理意义

4）掌握失速的现象和发生的原因

5）根据升力、阻力公式和对应的曲线解决飞行中的空气动力学问题

6）培养严谨、细心的职业素养，以及有效表达、协调和沟通的能力

Learning Guide

The aerodynamic performance of an aircraft cannot be measured solely from the perspective of lift or drag. It is necessary to combine the two to analyze the comparative relationship between lift and drag. The lift drag ratio refers to the ratio of lift to drag at the same angle of attack. The lift drag ratio is the ratio of lift coefficient to drag coefficient at the same angle of attack. Since the lift coefficient and drag coefficient change mainly with the angle of attack, the lift drag ratio also changes mainly with the angle of attack. In other words, the lift drag ratio has nothing to do with the density of air, flight speed and small wing area. Because these factors change, lift and drag both change in the same proportion without affecting the ratio of the two. A large lift to drag ratio indicates that the resistance is relatively small when obtaining the same lift. The higher the lift to drag ratio, the better the aerodynamic performance of the aircraft and the more advantageous it is for flight.

 课文

Lift/Drag Formulation

升力、阻力公式

The lift is formulated as

$$L=C_L \cdot \frac{1}{2}\rho v^2 \cdot S$$

The drag is formulated as

$$D=C_D \cdot \frac{1}{2}\rho v^2 \cdot S$$

Where lift and drag coefficient are C_L and C_D respectively; $\frac{1}{2}\rho v^2$ is the dynamic pressure; ρ is the density of the air; v is the flight speed; S is the area of the wing.

The lift and drag are directly proportional to the density of the air, the square of the flight speed, the lift and drag coefficient, and the area of the wing.

升力公式：

$$L=C_L \cdot \frac{1}{2}\rho v^2 \cdot S$$

阻力公式：

$$D=C_D \cdot \frac{1}{2}\rho v^2 \cdot S$$

式中，C_L 和 C_D 分别是升力系数和阻力系数；$\frac{1}{2}\rho v^2$ 是动压力；ρ 是空气密度；v 是飞行速度；S 是机翼面积。

升力和阻力分别与空气密度、飞行速度的平方、升力和阻力系数以及机翼面积成正比。

1. Density of the Air

1. 空气密度

The density of the air decreases with the increase of the altitude and the temperature. If an aircraft is going to take off at an airport with high altitude in hot weather, in order to achieve the lift required for taking off, it is necessary to increase the departure speed due to the low air density.

The low density of the air makes it difficult for the aircraft to accelerate. This affects the aircraft in high temperature and high altitude airports. The higher the altitude the lower the air density, the smaller the lift of the aircraft. In order to obtain the enough lift for flight, the flight speed of the aircraft must be increased.

空气密度随着海拔和温度的升高而降低。飞机如果在炎热天气下的高空机场起飞，为了达到起飞所需的升力，由于空气密度低，有必要提高起飞速度。

较低的空气密度使飞机难以加速。这会影响高温和高海拔机场的飞机。海拔越高，空气密度越低，飞机的升力越小。为了获得飞行所需的升力，必须提高飞机的飞行速度。

2. Square of the Flight Speed

2. 飞行速度的平方

With the increase of the flight speed, lift and drag will increase exponentially.

随着飞行速度的增加，升力和阻力呈指数增长。

3. Area of the Wings

3. 机翼的面积

When the area of the wings increases, the lift and drag increase as well.

The speed of the early aircraft is very slow, the large area of the wings is adopted to obtain the enough lift for flight. With the increase of the speed, the main problem to be resolved is how to reduce the drag. Therefore, the area of the wings tends to decrease. As we can see nowadays, the area of the wings in supersonic aircraft is very small.

当机翼面积增加时，升力和阻力也会增加。

早期的飞机飞行速度非常慢，通常依靠增加机翼面积获得飞行所需的升力。随着飞机速度的增加，主要解决的问题是如何减小阻力，而减小飞机机翼面积是减小阻力的方法之一。现在超声速飞机的机翼面积很小。

4. Lift and Drag Coefficient

4. 升力和阻力系数

(1) The lift is proportional to the lift coefficient, and the drag is proportional to the drag coefficient. Lift coefficient and drag coefficient are non-dimensional coefficient. When the flight speed doesn't exceed the limitation, both coefficients are only effected by the airfoil, shape and angle of attack of the wing.

(2) The thick airfoils and the airfoils whose maximum thickness point is at the front can accelerate the airflow rapidly through the upper wing's surface. That decreases the pressure,

leading to a large lift coefficient.

(3) The large camber airfoils and the airfoils whose maximum camber point is at the front can increase the lift coefficient. They are commonly used in low speed aircraft.

(4) Increasing the thickness and the camber of the airfoil will result in the large drag coefficient, leading to the large drag of the aircraft. Thin airfoils, the airfoils whose maximum thickness point is at the rear, or symmetrical thin airfoils without the camber are commonly adopted in high speed aircraft.

（1）升力与升力系数成正比，阻力与阻力系数成正比。升力系数和阻力系数是无量纲系数。当飞行速度不超过限制时，它们仅受机翼翼型、形状和迎角的影响。

（2）具有大厚度和最大厚度点靠前的翼型可以加速气流快速通过机翼上表面，压力降低，从而产生较大的升力系数。

（3）具有大弯度和最大弯度点靠前的翼型可以提高最大升力系数，通常用于低速飞机。

（4）增加机翼的厚度和弯度会增加阻力系数，从而增加飞机的飞行阻力。在高速飞机中，采用具有小厚度和最大厚度点靠后的薄翼型，或无弯度的对称薄翼型。

5. Lift Coefficient vs. AoA

5. 升力系数与迎角

When AoA is within the limitation, the lift coefficient is approximately linear with AoA. When AoA increases, the lift coefficient increases from negative value through zero to positive value, then to the maximum value. After that, it begins to decline. The angle of attack at the point where the lift coefficient is zero is called zero-lift AoA. For example, zero-lift AoA in Fig. 4-19 is $-5°$.

The zero-lift AoA is a small negative value for the asymmetric airfoils with a certain camber. These airfoils are used in most of the civil transport aircraft. If the AoA is less than the zero-lift AoA, the lift coefficient is negative, and the lift generated by the wing is negative. If the AoA is greater than the zero-lift AoA, the lift coefficient is positive, and the lift generated by the wing is positive.

The AoA corresponding to the maximum lift coefficient is called the critical angle of attack. The lift coefficient declines rapidly when the flying AoA exceeds the critical AoA.

当迎角在限制范围内时，升力系数与迎角近似线性。当迎角增加时，升力系数从负值增加到正值，并逐步达到最大值，然后开始下降。在升力系数为零时的迎角称为零升力迎角。图 4-19 中的零升力迎角为 $-5°$。

对于具有一定弯度的非对称翼型，零升力迎角是一个较小的负值，这些翼型用于大多数民用运输机。如果迎角小于零升力迎角，升力系数为负，机翼产生的升力为负。如果迎角大于零升力迎角，升力系数为正，机翼产生的升力为正。

对应最大升力系数的迎角称为临界迎角。当飞行迎角超过临界迎角时，升力系数迅速下降。

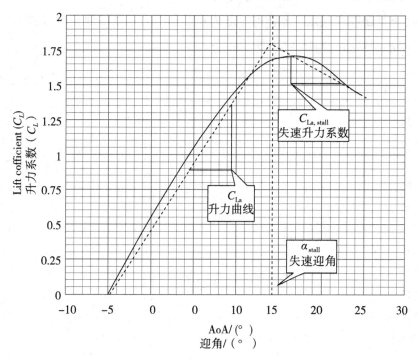

Fig. 4-19 Lift cofficient vs. AoA

图 4-19 升力系数与迎角

6. Drag Coefficient vs. AoA

6. 阻力系数与迎角

The drag coefficient is the smallest when AoA is equal to zero. The drag coefficient increases when AoA is greater or less than zero, as shown in Fig. 4-20.

当迎角等于零时，阻力系数最小。当迎角低于或高于零度时，阻力系数增加，如图 4-20 所示。

7. Lift Drag Ratio vs. AoA

7. 升阻比与迎角

(1) Lift drag ratio is the the ratio of lift coefficient to drag coefficient.

(2) When the flight speed is within certain ranges, both the lift coefficient and the drag one increase with the increasing of AoA, the lift drag ratio also increases, as shown in Fig. 4-20.

(3) The lift drag ratio begins to decline after it reaches the maximum at a certain AoA.

(4) The specific AoA when the lift drag ratio reaches the maximum is the most beneficial to the flight. At this time, the flight efficiency is the highest. Because, the smallest drag is generated

when the aircraft flying with the same lift at the specific AoA, the lift drag ratio is also called aerodynamic efficiency.

（1）升阻比是升力系数与阻力系数之比。

（2）当飞行速度在一定范围内时，升力系数和阻力系数随迎角的增加而增加，升阻比也随迎角增加而增加，如图 4-20 所示。

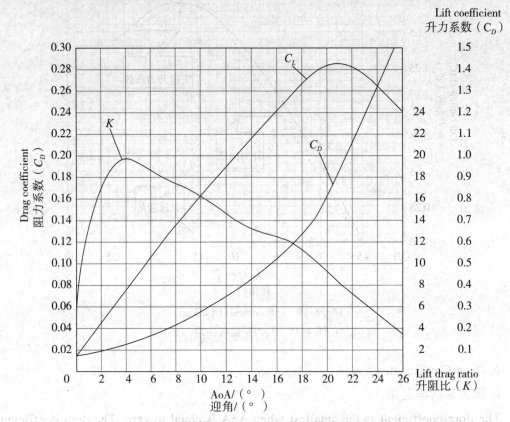

Fig. 4-20 Lift coefficient / Drag coefficient / Lift drag ratio vs. AoA

图 4-20 升力系数、阻力系数、升阻比与迎角的关系

（3）升阻比达到最大后，随着迎角的增加而减小。

（4）当升阻比达到最大时，飞机飞行最为有利，此时飞行效率最高。因为在这个迎角下，当飞机以相同的升力飞行时，产生的阻力最小，所以升阻比也被称为气动效率。

8. Stall

8. 失速

When AoA is greater than the critical AoA, the lift coefficient decreases and the drag coefficient increases rapidly, leading to the deterioration of aerodynamic performance in flight. This phenomenon is called stall.

The large AoA makes the separation of the boundary layers on the upper wing surface, leading to a large area of vortex, producing a large pressure drag. The appearance of vortex not only makes the lift and drag change rapidly, but also makes the aircraft speed and altitude decrease and nose down. In addition, because the separation is unstable and periodical, the lift changes randomly, resulting in the vibration of wing and tail, the decline of aircraft stability and maneuverability. It is difficult to operate the aircraft, and it is very dangerous for the aircraft to fly. The stall caused by excessive AoA may occur as long as the current AoA exceeds the critical AoA of the aircraft under any airspeed and flight conditions.

Usually, the flight state corresponding to the maximum lift coefficient and critical AoA will not be reached in flight. Before reaching this state, the phenomena of stall such as vibration and deterioration of stability have occurred because of the extended boundary layer separation region. In order to ensure flight safety and prevent stall, the lift coefficient must less than the maximum lift coefficient and the AoA is less than the critical AoA. Both values are safety values that can be reached but cannot be exceeded in flight.

当迎角大于临界迎角时，升力系数减小，阻力系数迅速增大，导致飞行气动性能恶化。这种现象称为失速。

大迎角导致机翼上表面的附面层分离，形成大面积涡流，从而产生大的压差阻力。涡流的出现不仅使升力和阻力迅速变化，还导致飞机速度、高度和机头下降。此外，分离不稳定且具有周期性，使升力不稳定，导致机翼和尾翼振动，飞机稳定性和机动性下降，飞机难以维持正常飞行。此时，飞机飞行非常危险。在任何空速和飞行姿态下，只要迎角超过飞机的临界迎角，就可能发生由过大迎角引起的飞机失速。

通常，在飞行中不会达到对应于最大升力系数和临界迎角的飞行状态。在达到该状态之前，由于附面层分离区域的扩大，出现了诸如振动和稳定性恶化等失速现象。为了确保飞行安全和防止飞机失速，规定了一个小于最大升力系数的升力系数值和一个小于临界迎角的迎角值。这两个值是飞行中可以达到但不能超过的安全值。

9. Stall Speed
9. 失速速度

The speed at which an aircraft reaches the critical AoA is called stall speed.

In the specific flight state, when the aircraft's mass increases, the required lift will be increased. However, the maximum lift coefficient of the aircraft is basically unchanged. The stall speed of the aircraft will increase, by increasing the flight speed.

In the takeoff and landing process, the lift increasing device can be adopted to improve the maximum lift coefficient, so as to reduce the stall speed of the aircraft, let the aircraft take off and land at a lower speed.

飞机达到临界迎角时的速度称为失速速度。

在同一飞行状态下，当飞机质量增加时，所需升力增加，但飞机的最大升力系数基本不变，只有提高飞行速度。这样，飞机的失速速度也就增加了。

在飞机起飞和着陆过程中，加装增升装置可以提高飞机的最大升力系数，从而降低飞机的失速速度，让飞机以较低的速度起飞和着陆。

Aerodynamic
Force Factors (1)

Aerodynamic
Force Factors (2)

Aerodynamic
Force Factors (3)

Aerodynamic
Force Factors (4)

Aerodynamic
Force Factors (5)

Aerodynamic
Force Factors (6)

Aerodynamic
Force Factors (7)

 New Words

formulate	['fɔːrmjuleɪt]	vt.	制定，确切表达
coefficient	[ˌkəʊɪ'fɪʃnt]	n.	系数
		adj.	共同作用的
respectively	[rɪ'spektɪvli]	adv.	分别地
proportional	[prə'pɔːrʃənl]	n.	比例，比例量
		adj.	成比例的
square	[skwer]	n.	平方，正方形
		adj.	方的，正方形的
		vt.	使成正方形
altitude	['æltɪtuːd]	n.	海拔高度，海拔
take off	['teɪk ɔːf]	v.	起飞
airport	['erpɔːrt]	n.	航空港，机场
weather	['weðər]	n.	天气，气象
achieve	[ə'tʃiːv]	v.	实现，完成
departure	[dɪ'pɒrtʃər]	n.	离开，出发
supersonic	[ˌsuːpər'sɒnɪk]	n.	超声波，超声速飞机
		adj.	超声速的

exponentially	[ˌɛkspəʊˈnɛnʃəli]	adv.	以指数方式
dimension	[daɪˈmenʃn]	n.	尺寸，规模，程度
Mach number	[mɒk ˈnʌmbər]	n.	马赫数
exceed	[ɪkˈsiːd]	vt.	超过
rapidly	[ˈræpɪdli]	adv.	迅速地
symmetrical	[sɪˈmetrɪkl]	adj.	对称的
linear	[ˈlɪniər]	adj.	线性的，直线的
decline	[dɪˈklaɪn]	n.	减少，下降
		v.	减少，下降
asymmetric	[ˌeɪsɪˈmetrɪk]	adj.	不对称的，不对等的
degree	[dɪˈgriː]	n.	度，程度，度数
range	[reɪndʒ]	n.	范围，射程
beyond	[bɪˈjɒnd]	n.	那边
		adv.	在另一边更远处，以远
		prep.	超过，超出
advantageous	[ˌædvənˈteɪdʒəs]	adj.	有利的，有好处的
efficiency	[ɪˈfɪʃnsi]	n.	效率，效能，功效
critical	[ˈkrɪtɪkl]	adj.	关键的，至关紧要的
critical angle of attack	[ˈkrɪtɪkl ˈæŋgl əv əˈtæk]		临界迎角
rapidly	[ˈræpɪdli]	adv.	迅速地
stalling	[ˈstɔːlɪŋ]	v.	失速
deterioration	[dɪˌtɪriəˌreɪʃən]	n.	变坏，退化
phenomenon	[fəˈnɒmɪnən]	n.	现象
vibration	[vaɪˈbreɪʃn]	n.	振动，震动，颤动，抖动
nose down	[nəʊz daʊn]		俯冲，机头向下
periodical	[ˌpɪəriˈɒdɪk(ə)l]	n.	期刊
		adj.	定期的，时常发生的
stability	[stəˈbɪləti]	n.	稳定性，稳定（性）
maneuverability	[məˌnuvərəˈbɪlɪti]	n.	机动性，可操纵性
excessive	[ɪkˈsesɪv]	v.	过分的，过度的
expansion	[ɪkˈspænʃn]	n.	膨胀，扩张
ensure	[ɪnˈʃʊr]	v.	确保，保证
landing	[ˈlændɪŋ]	n.	降落，着陆，登陆
		v.	落，降落，着陆

Q&A

The following questions are for you to answer to assess the learning outcomes.

(1) Please repeat the lift formula and the meaning of each parameter.

(2) Please repeat the drag formula and the meaning of each parameter.

(3) How does the air density change with altitude and temperature?

(4) How does the lift coefficient change with AoA?

(5) How does the drag coefficient change with AoA?

(6) How does the lift drag ratio change with AoA?

(7) Describe the definition of the stall speed.

 Extended Reading

What Happens When an Airplane Stalls?

1. What Is a Stall?

Put simply, a stall is a reduction of lift experienced by an aircraft. It occurs when the angle of attack of the wing is increased too much. This is known as the critical angle of attack and is typically around 15° (but there are variations).

In normal flight, the airflow over the shaped wings creates lift. The airfoil shape changes the airflow direction, and the downward deflection of the air causes an upward force (lift) to be exerted on the airfoil.

Increasing the angle of attack causes flow separation, where the air no longer flows cleanly over the upper surface of the wing. If this angle reaches the critical angle, the airflow is disrupted to the point where the lift generated begins to decrease.

It is the angle of attack of the wing exceeding its critical angle that causes a stall, but airspeed is also important. A "stall speed" will be defined for an aircraft, rather than an angle of attack. How does this work?

If an aircraft flies slower, it required a greater angle of attack to generate sufficient lift. If the speed decreases to a certain level, this angle will reach the critical angle. At this speed, an aircraft cannot climb without causing a stall. This speed is affected by several factors, including mass, altitude, and configuration, and different stall speeds are set based on this (such as a minimum speed in landing configuration with fully extended flaps).

2. What Happens in a Stall, and Why Is It Dangerous?

An uncorrected stall will cause the aircraft to fall. The first sign for a pilot is sluggish flight controls, which become much less responsive due to the changes in airflow, and possible buffeting. Pilots will train to recognize this, but this is more relevant in smaller aircraft that are flown manually.

An early stall is easily corrected by pushing the aircraft nose down to reduce the angle of attack. This is, of course, much more serious at low altitude when taking off or landing. If not corrected, the wing loses lift, and the aircraft will start to fall.

A spin is another situation that can occur. This occurs when the aircraft has sufficient yaw at the point of stall. In this situation, one wing stalls before the other, and the difference in lift causes the aircraft to roll. This is much harder for a pilot to recover from. Training is sometimes given on smaller aircraft as part of pilot training, but in general, the focus is on preventing a spin from ever happening. Commercial airliners are not designed or tested in this area.

3. Warning of a Stall

Any fixed-wing aircraft can stall. And all aircraft have warning systems to prevent, or alert pilots, to dangers. On a smaller, light aircraft, the most common method involves a simple flap on the leading wing edge, designed to activate a warning if the wing approaches its critical angle of attack.

Modern fly-by-wire aircraft will incorporate several systems to alert pilots of an approaching stall. This includes monitoring of speed and sensors to measure the angle of attack. Warnings can be given by alarm as well as mechanical "stick shakers" designed to give similar warnings to manual controls.

These sensors are part of the problem that previously grounded the Boeing 737 MAX aircraft. The Maneuvering Characteristics Augmentation System (MCAS) system takes data from the angle of attack sensors. Erroneous input from one of these sensors is thought to have led to the aircraft's nose being forced down in both tragic MAX accidents.

4. Stalls have Caused Many Accidents

Despite training and warning systems, stalls do still occur. At low speeds and low altitude during takeoff and landing, they can be disastrous, and unfortunately, a number of crashes have occurred. Some of the most notable include:

British European Airways Flight 548, June 1972. This is one of the most deadly crashes ever in the UK. A Trident aircraft stalled and hit the ground shortly after departure from Heathrow when the captain failed to maintain sufficient airspeed in the climb.

Air France Flight 447, June 2003. An Airbus A330 flying from Rio de Janeiro to Paris stalled at a high altitude (after the autopilot system was disabled due to airspeed measurement problems). The pilots failed to recover, and the aircraft descended to hit the ocean.

Turkish Airlines Flight 1951, February 2009. This involved a Boeing 737-800, which crashed on landing at Amsterdam. A failed radio altimeter causesd the engine power to be automatically reduced to idle, leading to a stall that pilots had no time to recover from.

任务 4　压力中心和气动中心
Task 4　Center of Pressure and Aerodynamic Center

Contents

1) Center of pressure
2) Aerodynamic center
3) CP vs. AC

Learning Outcomes

1) Master the definition of the center of pressure and aerodynamic center
2) Understand the role and physical characteristics of the center of pressure and aerodynamic center
3) Solve practical aerodynamic problems by the center of pressure and aerodynamic center

任务内容

1）压力中心
2）气动中心
3）压力中心和气动中心的区别与联系

任务目标

1）掌握压力中心和气动中心的定义
2）理解压力中心和气动中心的作用和物理意义
3）利用压力中心和气动中心的概念解决实际的空气动力学问题

If the moment generated by aerodynamic forces at a point on an object (such as a wing) is independent of the angle of attack of the object, then this point is called the aerodynamic center of the object. That is to say, the torque at a point does not vary with the angle of attack, and this point is the aerodynamic center.

 课文

1. Center of Pressure
1. 压力中心

The center of pressure on the airfoil is the point at which the aerodynamic force acts on the airfoil. When the air flows through the airfoil, the aerodynamic force is the distributed load on the airfoil surface. The resultant of these distributed loads is the aerodynamic force on the airfoil, and the point where the force acts on the airfoil is the center of pressure.

机翼的压力中心是作用在机翼上的气动载荷作用点。当空气流经机翼时，气动力是作用在机翼表面上的分布载荷。这些分布载荷的合力是机翼的气动力，合力的作用点是机翼的压力中心。

2. Aerodynamic Center
2. 气动中心

The wing aerodynamic center is the point of wing on which the change of aerodynamic lift is acting. When the angle of attack changes, the position of the pressure center of the wing changes, but the torque to the aerodynamic center remains unchanged. For example, because of gust disturbance, the aircraft's angle of attack increases, resulting in the increase of aerodynamic lift of the wing and the forward movement of the pressure center. With the concept of aerodynamic center, this change effect can be replaced by the effect of the original aerodynamic lift acting on the original pressure center and the aerodynamic lift increment caused by the change of angle of attack acting on the aerodynamic center.

机翼气动中心是机翼上气动升力变化的作用点。当迎角改变时，机翼压力中心的位置改变，但对气动中心的力矩保持不变。例如，由于阵风干扰，飞机的迎角增加，导致机翼的气动升力增加，压力中心向前移动。利用气动中心的概念，这种变化效应可以作用在原始压力中心上的原始气动升力和作用在气动中心上的迎角变化引起的气动升力增量的效应来代替。

3. CP vs. AC
3. 压力中心和气动中心的区别与联系

The center of pressure is the acting point of the aerodynamic force of the wing. The aerodynamic center is the action point of wing aerodynamic increment. The position of wing center of pressure moves forward and backward with the change of angle of attack. Within a certain range of angle of attack, when the angle of attack increases and the wing center of pressure moves forward, and when the angle of attack decreases and the wing center of pressure moves backward. In a low speed flight, the aerodynamic center of the wing does not change with the angle of attack.

When analyzing the effect of wing aerodynamic change on aircraft stability and maneuverability due to the change of angle of attack, the increment of aerodynamic lift can be applied to the aerodynamic center only. That is, only the influence of the increment of aerodynamic lift acting on the aerodynamic center would be concerned.

压力中心是机翼气动力的作用点。气动中心是机翼气动增量的作用点。机翼压力中心的位置随着迎角的变化而前后移动。在一定迎角范围内，当迎角增大时机翼压力中心向前移动，或当迎角减小时机翼压力重心向后移动。在低速飞行中，机翼的气动中心不随迎角变化。

分析由于迎角变化引起的机翼气动变化对飞机稳定性和机动性的影响时，气动升力增量只能应用于气动中心。也就是说，只考虑作用在气动中心上的气动升力增量的影响。

Center of Pressure and Aerodynamic Center (1)　　Center of Pressure and Aerodynamic Center (2)　　Center of Pressure and Aerodynamic Center (3)

 New Words

center of pressure	['sentər əv 'preʃər]		压力中心
aerodynamic center	[erəʊdaɪ'næmɪk 'sentər]		气动力中心，焦点，气动中心
distributed	[dɪ'strɪbjuːtɪd]	v.	使分布，分散
		adj.	分布的，分散的
load	[ləʊd]	n.	负载，负荷，装载量
		v.	承载，装载
torque	[tɔːrk]	n.	扭矩，转矩
gust	[gʌst]	n.	一阵强风，突风
		vi.	猛刮，劲吹

disturbance	[dɪ'stɜːrbəns]	n.	紊乱，乱流
increment	['ɪŋkrəmənt]	n.	增量，增加
analyze	['ænəlaɪz]	v.	分析，解析
concern	[kən'sɜːrn]	n.	关心，负责的事
		vt.	涉及，影响，牵涉

 Q&A

The following questions are for you to answer to assess the learning outcomes.

(1) Describe the definition of CP.

(2) Describe the definition of AC.

(3) Describe the variation of AC with the change of angle of attack.

 Extended Reading

Aerodynamic Center Does Not Move with Angle

Velocity produces a variation of pressure on the surface of the object. Integrating the pressure times the surface area around the body determines the aerodynamic force on the object. We can consider this force to act through the average location of the pressure on the surface of the object which we call the center of pressure in the same way that we call the average location of the mass of an object the center of gravity. In general, the pressure distribution around the object also imparts a torque, or moment, on the object. If a flying airfoil is not constrained in some way, it will flip as it moves through the air.

If we consider an airfoil at angle of attack, we can (theoretically) determine the pressure variation around the airfoil, and calculate the aerodynamic force and the center of pressure. But if we change the angle of attack, the pressure distribution changes and therefore the aerodynamic force and the location of the center of pressure also change. Since the pressure distribution changes with angle of attack, the torque created by this force also changes. So determining the aerodynamic behavior of an airfoil is very complicated if we use the center of pressure to analyze the forces.

For a single angle of attack, we can compute the moment about any point on the airfoil. The aerodynamic force will be the same, but the value of the moment depends on the point where that force is applied. It has been found both experimentally and theoretically that, if the aerodynamic force is applied at a location 1/4 chord back from the leading edge on most low speed airfoils, the magnitude of the moment is always the same, regardless of the angle of attack. Engineers call the location where the moment remains constant the aerodynamic center of the airfoil. Using the aerodynamic center as the location where the aerodynamic force is applied eliminates the problem of the movement of the center of pressure with angle of attack in aerodynamic analysis. (For supersonic airfoils, the aerodynamic center is nearer the 1/2 chord location.)

For symmetric airfoils, the aerodynamic moment about the AC is zero for all angles of attack. With camber, the moment is non−zero and constant for thin airfoils. For a positive cambered airfoil, the moment is negative and results in a counter−clockwise rotation of the airfoil. With camber, an angle of attack can be determined for which the airfoil produces no lift, but the moment is still present. This set of conditions is used experimentally to determine the aerodynamic moment which is then applied for all other flight conditions. For rectangular wings, the wing AC is the same as the airfoil AC. But for wings with some other planform (triangular, trapezoidal, compound, etc.) we have to find a mean aerodynamic center (MAC) which is the average for the whole wing. The computation of the MAC depends on the shape of the planform.

高速飞行
High Speed Flight

Contents

1) Air compressibility

2) Acceleration characteristics of high speed airflow

3) Shock and expansion waves

4) Critical Mach number and critical speed

5) Sound and thermal barriers

6) Aerodynamic configuration of high speed aircraft

学习内容

1）空气的可压缩性

2）高速气流的加速特性

3）激波和膨胀波

4）临界马赫数和临界速度

5）声障和热障

6）高速飞机的气动外形

任务 1　空气的可压缩性
Task 1　Air Compressibility

 Contents

1) Air compressibility

2) Mach number

3) Energy of air

Learning Outcomes

1) Understand the relationship between air compressibility and speed

2) Understand the relationship between air compressibility and sound speed

3) Master the definition and physical meaning of Mach number

4) Solve the aerodynamic problems in the high speed flight by using the air compressibility

5) Cultivate professional qualities of rigor, carefulness, and ability to express, coordinate, and communicate effectively

任务内容

1）空气的可压缩性

2）马赫数

3）空气能量

任务目标

1）理解空气可压缩性和速度的关系

2）理解空气可压缩性和声速的关系

3）掌握马赫数的定义和物理意义

4）运用空气可压缩性解决高速飞行中的空气动力学问题

5）培养严谨、细心的职业素养，以及有效表达、协调和沟通的能力

Learning Guide

Airflow characteristics refer to the relationship between the pressure, density, temperature, and thickness of the flow tube of the flowing air, as well as the mutual changes in airflow speed. In the process of changing the airflow speed from low to high, or from below the speed of sound to above the speed of sound, this relationship is also different. When the airflow speed exceeds the speed of sound, significant changes in pressure, density, and temperature occur due to strong compression or expansion of the air, resulting in some qualitative differences in airflow characteristics that are different from those at low speeds. For example, when the airflow increases in speed, the flow tube does not converge but expands; When the airflow decelerates, the shock wave phenomenon of sudden pressure rise will occur. The fundamental reason why the high speed airflow is so qualitatively different from the low beam airflow is that there is compressibility in the air.

 课文

1. Air Compressibility
1. 空气的可压缩性

The air compressibility is the main factor that makes high speed flight different from low speed flights. Low speed flights simplifies the problem by treating air as incompressible. However, in high speed flights, the compressibility must be considered as it causes some essential changes in airflow.

The compressibility of air in different parts of the atmosphere is different, and is shown by the difference of local sound speed. The flight speed indicates the degree of the change of local air pressure during the flight of the aircraft , while the sound speed indicates the difficulty of the local air being compressed during the flight (Fig. 5-1).

空气的可压缩性是高速飞行不同于低速飞行的主要原因。低速飞行将空气视为不可压缩的，以简化问题。然而，在高速飞行中，必须考虑空气的可压缩性，因为它会导致气流发生一些本质变化。

大气层中各处空气的可压缩性是不同的，而可压缩性的大小又可以通过局部声速的不同表现出来。飞行速度表示飞机飞行过程中造成空气局部压力变化程度的大小，而声速的大小表示飞行过程中局部空气被压缩的难易程度（图 5-1）。

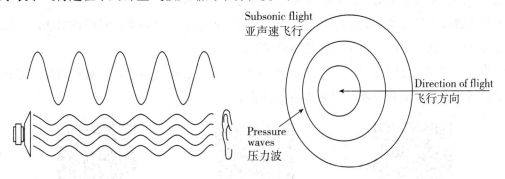

Fig. 5-1　Sound propagation
图 5-1　声音的传播

2. Mach Number
2. 马赫数

The Mach number of an aircraft is the ratio of the flight speed to the sound speed at the current flight altitude, which is a dimensionless value. The Mach number reflects not only the amount of pressure change exerted on the air during flight, but also the air compressibility. The greater the Mach number, the greater dagree of air being compressed by the aircraft, and the greater the impact on flight.

$$Ma = \frac{v}{a}$$

飞机飞行的马赫数是当前飞行高度下的飞行速度与声速之比，是一个无量纲的量。马赫数既反映了飞机飞行对空气施加的压力变化量的大小，也反映了空气可压缩性的大小。马赫数越大，飞机对空气的压缩程度越大，对飞行的影响就越大。

$$Ma = \frac{v}{a}$$

3. Energy of Air
3. 空气能量

In high speed flight, not only the degree of air compression increases, but also the temperature and internal energy of the air change with the flight speed. Therefore, the energy of air includes the conversion between mechanical energy and internal energy.

When the flight speed increases, the pressure, density, temperature of air and the sound speed decrease, and the Mach number increases. Part of the pressure energy and internal energy are transformed into dynamic energy, but the total energy of the system is conserved.

When the flight speed decreases, the pressure, density, temperature of air and the sound speed increase, and the Mach number decreases. Part of the dynamic energy is transformed into pressure energy and internal energy, but the total energy of the system is conserved.

在高速飞行中，不仅空气的压缩程度增大，空气的温度和内能也随着飞行速度的变化而变化。因此空气内能的转换包括机械能和内能之间的转换。

当飞机飞行速度增加时，空气的压力、密度、温度和声速降低，马赫数增加，部分压力能和内能转化为动能，但系统的总能量守恒。

当飞机飞行速度降低时，空气的压力、密度、温度和声速增加，马赫数减少，部分动态能量转化为压力能和内能，但系统的总能量守恒。

Air Compressibility (1)　　Air Compressibility (2)　　Air Compressibility (3)

simplify	['sɪmplɪfaɪ]	v.	简化，使简易
compressibility	[kəm,prɛsɪ'bɪlɪti]	n.	可压缩性，压缩常数
incompressible	[ɪnkəm'prɛsəbəl]	adj.	不能压缩的；坚硬的
cause	[kɔːz]	n.	原因，起因
		v.	导致，引起
essential	[ɪ'senʃl]	n.	要素，要点，必需品
		adj.	本质的，必不可少的
difficulty	['dɪfɪkəlti]	n.	困难，难题
current	['kɜːrənt]	n.	电流，水流
		adj.	现在的，当前的
dimensionless	[dɪ'mɛnʃənləs]	adj.	无量纲的
unit	['juːnɪt]	n.	单元，组件
impact	['ɪmpækt , ɪm'pækt]	n.	撞击，冲击力
		v.	冲击，撞击
compression	[kəm'prɛʃən]	n.	压缩，加压
conversion	[kən'vɜːrʒn]	n.	转变，转换
mechanical energy	[mə'kænɪkl 'enərdʒi]		机械能
internal energy	[ɪn'tɜːrnl 'enərdʒi]		内能
transform	[træns'fɔːrm]	n.	变换式
		v.	使改变，使转换

 Q&A

The following questions are for you to answer to assess the learning outcomes.

(1) Under what circumstances can air be regarded as incompressible?

(2) What parameters do we use to express air compressibility?

(3) Briefly describe the definition of Mach number.

 Extended Reading

Supersonic and Compressible Flows in Gas Turbines

Modern gas turbines commonly involve compressor and turbine blades that are moving so fast that the fluid flows over the turbine blades are locally supersonic. Density varies considerably in these flows so they are also considered to be compressible. Shock waves can form when these supersonic airflows are sufficiently decelerated. Shocks formed at blade leading edges or on blade surfaces can interact with other blades and shocks, and seriously affect the aerodynamic and structural performance the blade. It is possible to have supersonic airflows past blades near the outer diameter of a rotor, and subsonic flows near the inner diameter of the same rotor. These rotors are considered to be transonic in their operation. Very large aviation gas turbines can involve thrust levels exceeding 100,000 lb (1 lb=0.45359 kg). Two of these turbines are sufficient to carry over 350 passengers halfway around the world at high subsonic speeds.

任务 2　高速气流的加速特性
Task 2　Acceleration Characteristics of High Speed Airflow

 Contents

1) Acceleration characteristics of airflow

2) Laval nozzle

 Learning Outcomes

1) Master the acceleration characteristics of subsonic airflow

2) Master the acceleration characteristics of supersonic airflow

3) Use the acceleration characteristics of airflow to solve the aerodynamic problems in high speed flight

4) Cultivate professional qualities of rigor, carefulness, and ability to express, coordinate, and communicate effectively

 任务内容

1）气流加速特性

2）拉瓦尔喷管

 任务目标

1）掌握亚声速气流的加速特性

2）掌握超声速气流的加速特性

3）运用气流的加速特性解决高速飞行中的空气动力学问题

4）培养严谨、细心的职业素养，以及有效表达、协调和沟通的能力

 Learning Guide

Whether flying at low speeds or high speeds, the change of air velocity and pressure at both sides of the wing will cause the corresponding change of density of air. In the case of the large speed, the change of density of air caused by the change of airflow speed will cause additional changes in aerodynamics, and even change the airflow law.

 课文

1. Acceleration Characteristics of Airflow

1. 气流加速特性

(1) When $Ma < 0.4$, the air density changes are small and can be ignored. Therefore, the air density is considered to be constant. To accelerate the airflow, the cross section area of the flow tube must be reduced. Therefore, the low speed airflow is accelerated through the narrow part of the flow tube (Fig. 5-2).

(2) When $Ma = 1.0$, the cross section area of the flow tube remains unchanged.

(3) When $Ma > 1.0$, the air density decreases with the increase of Mach number. At this time, in order to keep the mass of airflow unchanged, the cross section area of the flow tube must be increased. That is, the supersonic airflow is accelerated by the expansion of the flow tube (Fig. 5-3).

（1）当 $Ma<0.4$ 时，空气密度变化很小，可以忽略不计，因此认为空气密度是恒定的。为了加速气流，必须减小流管的横截面积。低速气流是通过流管变细来实现加速的（图 5-2）。

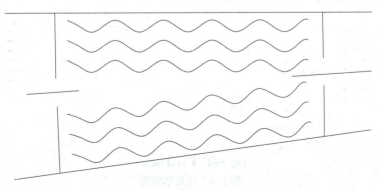

Fig. 5-2　Low speed airflow acceleration
图 5-2　低速气流加速

（2）当 Ma=1.0 时，流管的横截面积保持不变。

（3）当 Ma>1.0 时，空气密度随马赫数的增加而减小。此时，为了保持气流质量不变，必须增大流管的横截面积。也就是说，超声速流动是通过流管的扩张而加速的（图 5-3）。

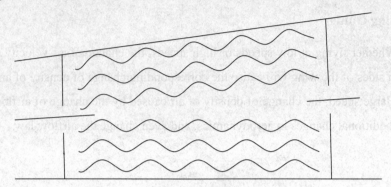

Fig. 5-3　High speed airflow acceleration

图 5-3　高速气流加速

2. Laval Nozzle
2. 拉瓦尔喷管

The narrowed flow tube can accelerate the subsonic airflow, but cannot obtain supersonic airflow. To accelerate the subsonic flow to supersonic speed, a flow tube that narrows and then expands is used. The subsonic airflow accelerates at the narrow part of the flow tube, reaches the sound speed at the narrowest part of the flow tube (the throat of the flow tube), and then continues to accelerate at the expansion part of the flow tube to become supersonic. A flow tube of this shape is called Laval nozzle, also known as supersonic nozzle (Fig. 5–4).

收缩的流管可以加速亚声速气流，但不能获得超声速气流。为了将亚声速气流加速到超声速，使用了先收缩后扩张的流管。亚声速气流在流管的收缩部分加速，在流管最细的部分（流管的喉部）达到声速，然后在流管扩张部分继续加速成为超声速气流。这种形状的流管称为拉瓦尔喷管，也称为超声速喷管（图 5-4）。

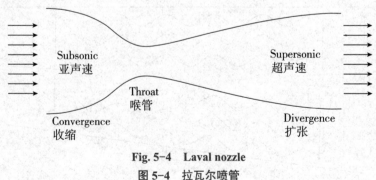

Fig. 5-4　Laval nozzle

图 5-4　拉瓦尔喷管

Acceleration Characteristics of High Speed Airflow (1)   Acceleration Characteristics of High Speed Airflow(2)

 New Words

ignore	[ɪɡ'nɔːr]	*v*	忽略，忽视，不予理睬
accelerate	[ək'seləreɪt]	*v.*	加快，加速
mass	[mæs]	*n.*	质量
		v.	集结，聚集
		adj.	大批的，广泛的
expansion	[ɪk'spænʃn]	*n.*	扩张，膨胀，扩展，扩大
narrow	['nærəʊ]	*n.*	狭窄部分，狭路，峡谷
		v.	变窄，缩小
		adj.	狭窄的，窄小的
supersonic	[ˌsuːpər'sɒnɪk]	*n.*	超声，超声速飞机
		adj.	超声速的
subsonic	[ˌsʌb'sɒnɪk]	*n.*	亚声速飞机
		adj.	亚声速的
throat	[θrəʊt]	*n.*	咽喉，喉咙，喉道
Laval	[lə'val]	*n.*	拉瓦尔
nozzle	['nɒzl]	*n.*	喷嘴，管口

 Q&A

The following questions are for you to answer to assess the learning outcomes.

(1) How to accelerate subsonic airflow?

(2) How to accelerate supersonic airflow?

(3) How to accelerate subsonic airflow to supersonic speed?

Compressible Aerodynamics

High speed aerodynamics, often called compressible aerodynamics, is a special branch of study of aeronautics. It is utilized by aircraft designers when designing aircraft capable of speeds approaching Mach 1 and above.

In the study of high speed aeronautics, the compressibility effects on air must be addressed. This flight regime is characterized by the Mach number, a special parameter named in honor of Ernst Mach, the late 19th century physicist who studied gas dynamics. Mach number is the ratio of the speed of the aircraft to the local speed of sound and determines the magnitude of many of the compressibility effects.

As an aircraft moves through the air, the air molecules near the aircraft are disturbed and move around the aircraft. The air molecules are pushed aside much like a boat creates a bow wave as it moves through the water. If the aircraft passes at a low speed, typically less than 250 m/s, the density of the air remains constant. But at higher speeds, some of the energy of the aircraft goes into compressing the air and locally changing the density of the air. The bigger and heavier the aircraft, the more air it displaces and the greater effect compression has on the aircraft.

This effect becomes more important as speed increases. Near and beyond the speed of sound, about 760 m/s (at sea level), sharp disturbances generate a shockwave that affects both the lift and drag of an aircraft and flow conditions downstream of the shockwave. The shockwave forms a cone of pressurized air molecules which move outward and rearward in all directions and extend to the ground. The sharp release of the pressure, after the buildup by the shockwave, is heard as the sonic boom.

Listed below are a range of conditions that are encountered by aircraft as their designed speed increases.

Subsonic conditions occur for Mach numbers less than 1 (100−350 m/s). For the lowest subsonic conditions, compressibility can be ignored.

As the speed of the object approaches the speed of sound, the flight Mach number is nearly equal to one, $Ma = 1$ (350−760 m/s), and the flow is said to be transonic. At some locations on the object, the local speed of air exceeds the speed of sound. Compressibility effects are the most important in transonic flows and lead to the early belief in a sound barrier. Flight faster than sound was thought to be impossible. In fact, the sound barrier was only an increase in the drag near sonic conditions because of compressibility effects. Because of the high drag associated with compressibility effects, aircraft are not operated in cruise conditions near Mach 1.

Supersonic conditions occur for numbers greater than Mach 1, but less then Mach 3 (760–2,280 m/s). Compressibility effects of gas are important in the design of supersonic aircraft because of the shockwaves that are generated by the surface of the object. For high supersonic speeds, between Mach 3 and Mach 5 (2,280–3,600 m/s), aerodynamic heating becomes a very important factor in aircraft design.

For speeds greater than Mach 5, the flow is said to be hypersonic. At these speeds, some of the energy of the object now goes into exciting the chemical bonds which hold together the nitrogen and oxygen molecules of the air. At hypersonic speeds, the chemistry of the air must be considered when determining forces on the object. When the Space Shuttle re-enters the atmosphere at high hypersonic speeds, close to Mach 25, the heated air becomes an ionized plasma of gas, and the spacecraft must be insulated from the extremely high temperatures.

任务 3　激波和膨胀波
Task 3　Shock Waves and Expansion Waves

Contents

1) Mach cone

2) Mach angle

3) Shock waves

4) Direct shock wave

5) Normal shock wave

6) Expansion waves

Learning Outcomes

1) Understand the generation conditions and significance of Mach cones

2) Master the generation conditions of shock waves

3) Master the changes of physical parameters before and after shock waves

4) Master the generation conditions of expansion waves

5) Master the changes of parameters before and after expansion waves

6) Use the concepts of shock and expansion waves to solve the aerodynamic problems

in high speed flights

7) Cultivate professional qualities of rigor, carefulness, and ability to express, coordinate, and communicate effectively

任务内容

1）马赫锥
2）马赫角
3）激波
4）波振面
5）正激波
6）膨胀波

任务目标

1）理解马赫锥的产生条件和意义
2）掌握激波的产生条件
3）掌握激波前后物理参数的变化情况
4）掌握膨胀波的产生条件
5）掌握膨胀波前后物理参数的变化情况
6）运用激波和膨胀波的概念解决高速飞行中的空气动力学问题
7）培养严谨、细心的职业素养，以及有效表达、协调和沟通的能力

Learning Guide

When an aircraft flies at supersonic speed in the air, waves occur.

课文

1. Mach Cone
1. 马赫锥

When the aircraft is flying in the air, tiny disturbances in the air propagates outward at the speed of sound, disturbing the surrounding air. If the speed of the aircraft is greater than the speed of sound, the aircraft will lead the disturbance wave to run in front of the disturbance wave. The tangent to each disturbance wave surface is made through the aircraft nose. The cone formed by the tangents is called Mach cone, as shown in Fig. 5–5.

飞机在空中飞行时，空气中的小扰动以声速向外传播，扰动周围的空气。如果飞机的

速度大于声速，飞机将领先扰动波而在扰动波前面飞行。每个扰动波面的切线通过飞机机头。由切线形成的圆锥称为马赫锥，如图 5-5 所示。

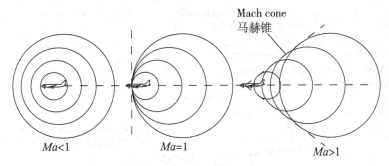

Fig. 5-5　Mach cone

图 5-5　马赫锥

2. Mach Angle
2. 马赫角

Mach angle is a function related to Mach number.

$$u=\arcsin \frac{1}{Ma}$$

Mach cone appears only when $Ma \geqslant 1$. The Mach angle is only related to the Mach number. The larger the Mach number is, the smaller the Mach angle and the range of disturbed area are. Mach cone is the interface between disturbed area and undisturbed area in supersonic flights.

马赫角是一个与马赫数有关的函数。

$$u=\arcsin \frac{1}{Ma}$$

马赫锥仅在 $Ma \geqslant 1$ 时出现。马赫角仅与马赫数有关。马赫数越大，马赫角越小，扰动区域范围越小。马赫锥是超声速飞行中扰动区和未扰动区之间的界面。

3. Shock Wave
3. 激波

A shock wave is an intense disturbance wave (compression wave) formed by intense compression when supersonic airflow flows through the surface of an object with internal folding angle, or when it flows around a large obstacle (Fig. 5-6).

With the increase of Mach number, the shock wave is gradually generated. The resistance of airflow through shock wave is called wave drag. After the airflow passes through the shock wave, the air parameters change dramatically. The air speed decreases, the temperature, pressure and density increase. The propagation speed of shock wave in air is greater than that of sound. The greater the intensity of shock wave is, the faster it propagates.

A surface formed by connecting points with the same vibration phase in a medium is called a wavefront.

激波是超声速气流流过带有内凹角的物体表面，或绕过较大的障碍物时，受到强烈压缩而形成的强扰动波（压缩波）（图 5-6）。

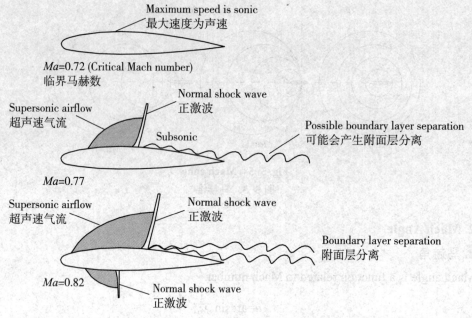

Fig. 5-6　Generation process of shock waves

图 5-6　激波产生过程

随着马赫数的增加，激波逐渐产生。气流通过激波时受到的阻力称为激波阻力。气流通过激波后，空气参数发生显著变化：空气速度降低，温度、压力和密度增加。激波在空气中的传播速度大于声速。激波强度越大，其传播速度越快。

同一时刻介质中振动相位相同的点联成的面称为波阵面。

4. Normal Shock Wave

4. 正激波

A normal shock wave is the wavefront of a shock wave that is perpendicular to the incoming airflow.

正激波是指与来流垂直的激波的波阵面。

5. Expansion Wave

5. 膨胀波

When the supersonic airflow flows over the object surface with external folding angle, or when the flow tube through which it flows becomes wider, the airflow speed increases and the pressure decreases. This causes the expansion and acceleration of the airflow, which is called the expansion wave (Fig. 5-7).

The airflow gradually accelerates and depressurizes through wave surfaces, and finally generates a higher speed airflow which flows along the surface of the object.

Supersonic airflow is compressed and decelerated by shock waves, and expanded and accelerated by expansion waves.

　　当超声速气流流过带有外凸角的物体表面时，或流过的流管变宽时，气流速度增加，压力降低。这导致气流膨胀和加速，称为膨胀波（图 5-7）。

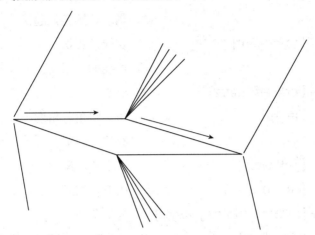

Fig. 5-7　Generation process of expansion waves
图 5-7　膨胀波的产生过程

　　气流通过多个波面逐渐加速和减压，最终产生更高速的气流，沿物体表面流动。超声速气流经过激波压缩并减速，经过膨胀波膨胀并加速。

| Shock Waves and Expansion Waves (1) | Shock Waves and Expansion Waves (2) | Shock Waves and Expansion Waves (3) |

 New Words

cone	[kəʊn]	n.	圆锥，圆锥体
		v.	使成锥形
Mach cone	[mɒk kəʊn]	n.	马赫锥，扰动锥
disturbance	[dɪ'stɜːrbəns]	n.	障碍，紊乱
propagate	['prɒpəgeɪt]	v.	传播
outward	['aʊtwərd]	n.	外表，外部
		adj.	表面的，外表的
		adv.	向外，朝外

surrounding	[sə'raʊndɪŋ]	n.	环境，周围的事物
		v.	围绕，环绕
		adj.	周围的，附近的
wave	[weɪv]	n.	波动，波浪
		v.	飘动，摇晃，起伏
tangent	['tændʒənt]	n.	切线，正切
		adj.	切线的，正切的
aircraft nose	['erkræft nəʊz]		机头
angle	['æŋgl]	n.	角，斜角，角度
		v.	斜移，斜置
shock wave	['ʃɒk weɪv]	n.	冲击波，激波
compression	[kəm'prɛʃən]	n.	压缩，浓缩
internal folding angle	[ɪn'tɜːnl 'fəʊldɪŋ 'æŋgl]		内折角
gradually	['grædʒuəli]	adv.	逐步地，逐渐地，渐进地
wave drag	[weɪv dræg]		波阻
dramatically	[drə'mætɪkli]	adv.	显著地
parameter	[pə'ræmɪtər]	n.	参数
intense	[ɪn'tens]	adj.	强烈的，激烈的
intensity	[ɪn'tensəti]	n.	强烈，强度
expansion wave	[ɪk'spænʃn weɪv]		膨胀波
external folding angle	[ɪk'stɜːnl 'fəʊldɪŋ 'æŋgl]		外折角
depressurize	[diː'preʃəraɪz]	v.	减压，降压
compress	[kəm'pres]	n.	压缩，压紧
		v.	压紧，精简，浓缩，压缩
expand	[ɪk'spænd]	v.	扩大，增加

 Q&A

The following questions are for you to answer to assess the learning outcomes.

(1) Describe the definition of Mach cone.

(2) Under what circumstances will Mach cone appear?

(3) Describe the definition of shock wave.

(4) How do air parameters change after airflow passes through shock waves?

(5) Describe the definition of expansion wave.

(6) How do air parameters change after airflow passes through expansion wave?

Shock Waves

Sound coming from an airplane is the result of the air being disturbed as the airplane moves through it, and the resulting pressure waves that radiate out from the source of the disturbance. For a slow moving airplane, the pressure waves travel out ahead of the airplane, traveling at the speed of sound. When the speed of the airplane reaches the speed of sound, however, the pressure waves (sound energy) cannot get away from the airplane. At this point the sound energy starts to pile up, initially on the top of the wing, and eventually attaching itself to the wing leading and trailing edges. This piling up of sound energy is called a shock wave. If the shock waves reach the ground, and cross the path of a person, they will be heard as a sonic boom. Fig. 5-8 shows a wing in slow speed flight, with many disturbances on the wing generating sound pressure waves that are radiating outward. Fig. 5-8(b) is the wing of an airplane in supersonic flight, with the sound pressure waves piling up toward the wing leading edge.

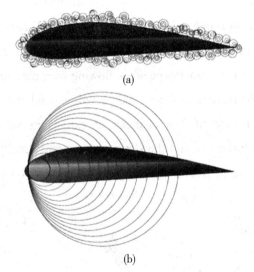

(a)

(b)

Fig. 5-8　Sound energy in subsonic and supersonic flight
图 5-8　亚声速和超声速飞行中的声能

1. Normal Shock Wave

When an airplane is in transonic flight, the shock wave that forms on top of the wing, and eventually on the bottom of the wing, is called a normal shock wave. If the leading edge of the wing is blunted, instead of being rounded or sharp, a normal shock wave will also form in front of the wing during supersonic flight. Normal shock waves form perpendicular to the airstream. The velocity of the air behind a normal shock wave is subsonic, and the static pressure and density of the air are higher. Fig. 5-9 shows a normal shock wave forming on the top of a wing.

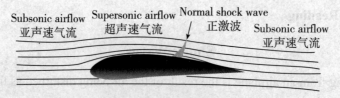

Fig. 5–9   Normal shock wave

图 5–9   正激波

2. Oblique Shock Wave

An airplane that is designed to fly supersonic will have very sharp edged surfaces, in order to have the least amount of drag. When the airplane is in supersonic flight, the sharp leading edge and trailing edge of the wing will have shock waves attach to them. These shock waves are known as oblique shock waves. Behind an oblique shock wave the velocity of the air is lower, but still supersonic, and the static pressure and density are higher. Fig. 5–10 shows an oblique shock wave on the leading and trailing edges of a supersonic airfoil.

3. Expansion Wave

Earlier in the discussion of high speed aerodynamics, it was stated that air at the supersonic speed acts like a compressible fluid. For this reason, supersonic air, when given the opportunity, wants to expand outward. When supersonic air is flowing over the top of a wing, and the wing surface turns away from the direction of flow, the air will expand and follow the new direction. At the point where the direction of flow changes, an expansion wave will occur. Behind the expansion wave the speed increases, and the static pressure and density decrease. An expansion wave is not a shock wave. Fig. 5–10 also shows an expansion wave on a supersonic airfoil.

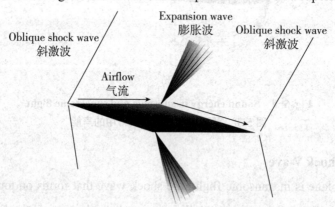

Fig. 5–10   Subsonic airfoils with oblique shock waves and expansion waves

图 5–10   具有斜激波和膨胀波的亚声波翼型

任务 4 临界马赫数和临界速度
Task 4 Critical Mach Number and Critical Speed

Contents

1) Local Mach number

2) Critical Mach number and critical speed

Learning Outcomes

1) Master the definition of critical Mach number

2) Use the concept of critical Mach number to solve the aerodynamic problems in high speed flight

3) Cultivate professional qualities of rigor, carefulness, and ability to express, coordinate, and communicate effectively

任务内容

1）局部马赫数

2）临界马赫数和临界速度

任务目标

1）掌握临界马赫数的定义

2）运用临界马赫数的概念解决高速飞行中的空气动力学问题

3）培养严谨、细心的职业素养，以及有效表达、协调和沟通的能力

Learning Guide

The speed of disturbance wave propagation is called wave speed. If the pressure difference before and after a strong disturbance wave is large, the wave speed is large. However, as the disturbance wave propagates forward, the pressure difference before and after the wave surface decreases, and its wave speed also slows down accordingly. The pressure difference between the front and back of a weakly perturbed wave is very small. Its propagation speed is the sound speed. The speed of sound is related to the medium through which sound waves propagate. The harder the medium is to compress, the higher the speed of sound. For example, the speed of sound propagates faster in metal than in water. The altitude decreases, the temperature is high, the air is not easily compressed, and the speed of sound is fast. Mach number is a very important parameter in aerodynamics, which can be used as a scale to divide the air speed, as a sign of the strength of air compressibility, and determine the propagation range of disturbance waves.

 课文

1. Local Mach Number
1. 局部马赫数

When the aircraft is flying, the airflow speed through the wing surface is not the same to the flight speed of the aircraft. For positive angle of attack, the airflow through the upper wing surface is accelerated, and its flow speed reaches the maximum at the lowest pressure point of the airfoil.

If each point in the flow field has different speeds, then the ratio of the flight speed to the sound speed at a point is called the local Mach number (Fig. 5-11).

飞机飞行时，流过机翼表面的气流速度与飞机的飞行速度并不相同。在正迎角情况下，流过机翼上表面的气流被加速，其流速在翼型的最低压力点达到最大。

如果流场中的各点速度不同，那么某一点的流速与该点声速的比值称为局部马赫数（图 5-11）。

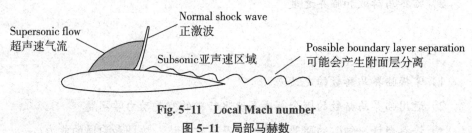

Fig. 5-11   Local Mach number
图 5-11   局部马赫数

2. Critical Mach Number and Critical Speed
2. 临界马赫数和临界速度

Considering the compressibility of air, if the airflow rate increases, the static pressure, temperature and density are going to decrease. Thus, the temperature and sound speed at the maximum speed point are also the lowest. Therefore, the local Mach number at this point is the largest in the flow field.

With the increase of aircraft flight speed, the local airflow speed at the maximum speed point is higher, the local sound speed is lower, and the local Mach number is getting higher. The local airflow speed at this point may reach sound speed, when the flight speed of the aircraft has not reached the sound speed of the flight altitude. That is, the Mach number is less than 1, but at the maximum speed point, the local Mach number has reached Mach 1. At this time, the Mach number of the aircraft is called the critical Mach number, and the speed of the aircraft is called the critical speed.

Critical Mach Number and Critical Speed (1)

考虑到空气的可压缩性，如果空气流量增加，静压、温度和密度将降低。这样，最高速度点的温度和声速也是最低的。因此，此处的局部马赫数在流场中最大。

Critical Mach Number and Critical Speed (2)

随着飞机飞行速度的增加，最大速度点的局部气流速度越高，局部声速减小，局部马赫数增大。当飞机的飞行速度尚未达到飞行高度的声速时，此时的局部气流速度可能达到声速，即马赫数小于1，但在最大速度点，局部马赫数已达到1。此时，飞机飞行的马赫数称为临界马赫数，飞机飞行的速度称为临界速度。

 New Words

local	['ləʊkl]	n. 当地人，附近
		adj. 地方的，局部的
local Mach number	['ləʊk(ə)l mæk 'nʌmbə(r)]	局部马赫数
flow field	[fləʊ fiːld]	流场
critical Mach number	['krɪtɪk(ə)l mæk 'nʌmbə(r)]	临界马赫数
critical speed	['krɪtɪkl spiːd]	临界速度

 Q&A

The following questions are for you to answer to assess the learning outcomes.

(1) Describe the characteristics of the maximum speed point of the wing.

(2) Describe the definition of the critical speed.

 Extended Reading

The Significance of the Critical Mach Number in Aerodynamics

1. Key Takeaways

The Mach number provides a comparison between fluid flow rate and the speed of sound. This metric is important in aerodynamics, as certain forces will increase as an aircraft approaches a critical Mach number.

The critical Mach number can be determined from fluid flow simulation along the body of an aircraft.

Although aircraft and fluid flow with regard to aircraft are complex systems, the primary forces and principles involved in studying aerodynamics are rather simple. We often discuss aerodynamics and aircraft in the context of low relative airflow speeds, where the craft is traveling below the speed of sound. The ratio of airflow speed to the speed of sound is known as the Mach number, and this will define some important characteristics of fluid flow behavior along the body of an aircraft.

During flight, as an aircraft accelerates and eventually reaches the Mach number, an approach to the speed of sound will have important implications on the aircraft's motion and total energy. It can also affect the ability of an aircraft to maintain its flight trajectory without experiencing stall, excessive drag, or loss of control due to buffeting. In this article, we'll examine what happens when an aircraft approaches the critical Mach number as well as the resulting effects once the Mach number is exceeded.

2. What Is the Critical Mach Number?

When airflow occurs across an aircraft during flight, it is not uniform along the airframe. When airflow over any portion of the aircraft approaches the speed of sound, flight characteristics will begin to change, which we will discuss below.

There is a critical Mach number, which will depend on the slowest part of the local flow speed across the body of an aircraft. Specifically, the critical Mach number is the slowest free stream Mach number at which the airflow along any other area of the aircraft reaches $Ma = 1$. In other words, the moment at which the fastest portion of the airflow equals the speed of sound, then the flow and the craft speed have reached the critical Mach number.

3. What Happens at the Critical Mach Number?

On the approach to the critical Mach number, the airflow exerts oscillatory (vibration) forces on the structure of the aircraft. This is known as buffeting, which occurs as $Ma = 1$ is approached, which is felt by the pilot as heavy vibrations. Once the speed of sound is exceeded for the entire flow, flight will feel smooth again. In terms of airflow, a shockwave occurs once the critical Mach number is exceeded, which will be heard by any surrounding observers.

Visually, this is examined by looking at spherical wavefronts produced by the moving aircraft. The aircraft will cause air compression along the leading edge of the airframe, and this air compression will then cause the emission of spherical acoustic waves around the front edge of the aircraft. At the front edge of the craft's nose cone, these air fronts pile up and eventually produce high compression along the front face of the aircraft when the Mach number reaches the limit of *Ma* = 1 in this region.

This point at which the airflow around the craft approaches the speed of sound, as well as other fluid flow effects that result from such a shockwave event, will depend on the shape and speed of the aircraft. How fluid flow along an aircraft is affected by its shape, and the forces exerted on the aircraft when approaching the critical Mach number, can be examined with CFD simulations.

任务 5　声障和热障
Task 5　Sound and Thermal Barriers

 Contents

1) Sound barrier

2) Sonic boom

3) Thermal barrier

4) Problems with thermal barriers

5) Solutions to thermal barrier problems

 Learning Outcomes

1) Understand the causes and phenomena of sound barrier

2) Understand the causes and phenomena of thermal barrier

3) Use the concepts of sound barrier and thermal barrier to solve the aerodynamic problems in high speed flight

4) Cultivate professional qualities of rigor, carefulness, and ability to express, coordinate, and communicate effectively

 任务内容

1）声障

2）声爆

3）热障

4）热障带来的问题

5）热障问题的解决方法

 任务目标

1）理解声障的产生原因和现象

2）理解热障的产生原因和现象

3）运用声障和热障的概念解决高速飞行中的空气动力学问题

4）培养严谨、细心的职业素养，以及有效表达、协调和沟通的能力

 Learning Guide

A series of abnormal phenomena caused by aircraft flying at speeds close to the speed of sound, such as a sharp increase in aircraft resistance, a decrease in lift, a decrease in propeller efficiency, strong body vibration, and control failure. In addition, when the aircraft is flying at supersonic speeds, its own materials will also undergo shape changes due to temperature rise, affecting flight.

课文

1. Sound Barrier

1. 声障

Once a subsonic aircraft approaches the critical Mach number, in addition to a sudden increase in drag, which makes it difficult for the aircraft to accelerate, the lift will also drop suddenly, resulting in aircraft stall. There will also be automatic head down, aircraft vibration, reduced control efficiency and automatic roll, which will make the aircraft out of control and even cause serious flight accidents. It is impossible to overcome these phenomena for transonic flight, even the power or thrust of subsonic aircraft engine is increased. These phenomena are called sound barrier.

一旦亚声速飞机接近临界马赫数，除阻力突然增大，使飞机难以加速外，升力会骤然下降，导致飞机失速；还会出现自动低头、飞机抖振、操纵效率降低和自动滚转，这将使飞机失去控制，甚至造成严重的飞行事故。即使增加亚声速飞机发动机的功率或推力，也不可能克服这些现象进行跨声速飞行。这些现象称为声障。

2. Sonic Boom

2. 声爆

The sonic boom is the sound waves generated by an aircraft flying at supersonic speed and transmitted to the ground to form an explosion sound (Fig. 5-12).

声爆是指飞机以超声速飞行时产生的声波传播到地面，形成爆炸声（图 5-12）。

Fig. 5-12　Sonic boom
图 5-12　声爆

3. Thermal Barrier

3. 热障

When the air flows through the aircraft, due to the boundary layer drag, the temperature increases, and the dynamic energy of the airflow is converted into thermal energy to heat the aircraft surface, which is aerodynamic heating. When flying at the subsonic speed, the heat generated by friction drag is less and cools in the air quickly. The surface temperature of the aircraft does not increase too much. However, the problem of aerodynamic heating becomes more serious when the aircraft flies at supersonic speed in the air. These phenomena are called thermal barrier.

当空气流经飞机时，由于附面层阻力，温度升高，气流的动能转化为热能而加热机体表面，这就是空气动力加热。当以亚声速飞行时，摩擦阻力产生的热量较少，并很快在空气中散失。飞机表面温度增加不多。当飞机在空中以超声速飞行时，空气动力加热问题会变得更加严重。这些现象称为热障。

1) Problems with Thermal Barriers

1）热障带来的问题

When cruising at the supersonic speed for a long time, aerodynamic heating will not only increase the aircraft surface temperature, but also heat the body structure and raise the cabin temperature, which will bring many problems to the flight of the aircraft.

(1) The cabin temperature is too high for the cabin personnel, the radio, aviation instruments and other airborne equipment cannot work normally.

(2) The temperature will exceed the limit of some non-metallic materials on the aircraft,

such as windshield, plexiglass and rubber for sealing. They will not work properly or even be completely damaged.

(3) The temperature of the aircraft reaches more than 200 ℃, which greatly reduces the mechanical properties of the aluminum alloy, the main structural parts material of the aircraft, reduces the strength and rigidity of it, and cannot meet the aircraft design requirements and thus the aircraft cannot fly normally.

当长时间以超声速巡航时，气动加热不仅会增加飞机表面温度，还会加热机体结构并使机舱温度升高，这将给飞机的飞行带来许多问题。

（1）对机舱人员来说温度会高得难以忍受，无线电、航空仪表等机载设备也无法正常工作。

（2）温度将超过飞机上某些非金属材料的极限工作温度，如挡风玻璃、有机玻璃、密封橡胶。它们将无法正常工作，甚至完全损坏。

（3）飞机的温度达到200 ℃以上，大大降低了飞机主要结构部件所使用的铝合金材料的机械性能，降低了飞机的强度和刚度，使其无法满足飞机的设计要求，因此无法正常飞行。

2) Solutions to Thermal Barrier Problems

2）热障问题的解决方法

The use of high temperature resistant materials such as titanium alloy and heat resistant alloy steel can improve the working temperature of the aircraft. But the aircraft speed also depends on the development of new structural materials such as advanced composite materials, and new process methods of these kinds of materials.

虽然使用耐高温材料，如钛合金和耐热合金钢，可以提高飞机的工作温度，但飞机飞行速度的进一步提高还取决于先进的复合材料等新型结构材料的研制和针对这些新型结构材料新加工工艺方法的开发。

Sound and Thermal Barriers (1)

Sound and Thermal Barriers (2)

Sound and Thermal Barriers (3)

Sound and Thermal Barriers (4)

 **New Words**

sound barrier	[saʊnd ˈbæriər]		声障
approach	[əˈprəʊtʃ]	*n.*	方法
		v.	靠近，接近
automatic	[ˌɔːtəˈmætɪk]	*n.*	自动手枪（或步枪；自动变速汽车；自动换挡汽车）
		adj.	自动的
head down	[hed daʊn]		飞机低头
control efficiency	[kənˈtrəʊl ɪˈfɪʃnsi]		控制效率
roll	[rəʊl]	*n.*	滚转
		v.	滚转
accident	[ˈæksədənt]	*n.*	事故，意外
impossible	[ɪmˈpɒsəbl]	*n.*	不可能，不可能的事
		adj.	不可能的
overcome	[ˌəʊvərˈkʌm]	*v.*	克服，解决
transonic	[trænˈsɒnɪk]	*n.*	跨声速
		adj.	跨声速的，超声速的
thermal barrier	[ˈθɜːrml ˈbæriər]		热障，温障
explosion	[ɪkˈspləʊʒn]	*n.*	爆炸，爆破
sonic boom	[ˌsɒnɪk ˈbuːm]	*n.*	声爆
aerodynamic heating	[ˌerəʊdaɪˈnæmɪk ˈhiːtɪŋ]		气动加热
cool	[kuːl]	*n.*	凉气，凉快
		v.	变凉，冷却
		adj.	凉的，凉爽的
serious	[ˈsɪriəs]	*adj.*	严重的，有危险的
structure	[ˈstrʌktʃər]	*n.*	结构，构造
		v.	使形成体系
cabin	[ˈkæbɪn]	*n.*	（飞机的）客舱
cabin personnel			客舱人员
radio	[ˈreɪdiəʊ]	*n.*	无线电广播
		v.	（用无线电）发送，传送
aviation	[ˌeɪviˈeɪʃn]	*n.*	航空
instrument	[ˈɪnstrəmənt]	*n.*	仪器，仪表
airborne	[ˈerbɔːrn]	*adj.*	空运的，飞机上的

metallic	[məˈtælɪk]	adj.	金属的，金属般的
material	[məˈtɪriəl]	n.	材料，原料
windshield	[ˈwɪndʃiːld]	n.	挡风玻璃，风挡
plexi glass	[ˈpleksi glɑːs]		有机玻璃
rubber	[ˈrʌbər]	n.	橡胶，橡皮
		adj.	橡胶制成的
completely	[kəmˈpliːtli]	adv.	彻底地，完全地，完整地
damage	[ˈdæmɪdʒ]	n.	损坏，破坏
		v.	毁坏，破坏
mechanical	[məˈkænɪkl]	adj.	机械驱动的，机械的
property	[ˈprɒpərti]	n.	性质，属性，财产
aluminum	[əˈlumənəm]	n.	铝
alloy	[ˈælɔɪ, əˈlɔɪ]	n.	合金
		v.	把……铸成合金
strength	[streŋθ]	n.	强度
rigidity	[rɪˈdʒɪdəti]	n.	刚性
design	[dɪˈzaɪn]	n.	设计，布局，安排
		v.	设计，制图
titanium	[tɪˈteɪniəm]	n.	钛
steel	[stiːl]	n.	钢
composite material	[kɒmˈpɒzɪt məˈtɪriəl]		复合材料
process	[ˈprəuses, prəˈses]	n.	过程，进程，步骤，流程
		v.	处理，加工

 Q&A

The following questions are for you to answer to assess the learning outcomes.

1) Describe the definition of sound barrier.

2) Describe the definition of sound boom.

3) Describe the definition of thermal barrier.

任务 6　高速飞机的气动特性
Task 6　Aerodynamic Configuration of High Speed Aircraft

Contents

1) Thin airfoil

2) Thin airfoil—laminar airfoil

3) Thin airfoil—supercritical airfoil

4) Swept wing

5) Disadvantages of swept wing

6) Low aspect ratio wing

7) Disadvantages of low aspect ratio wing

8) Vortex generator

9) Advantages of vortex generators

Learning Outcomes

1) Understand the aerodynamic configuration of high speed aircraft

2) Master the working principle of the aerodynamic configuration of high speed aircraft

3) Use the concept of the aerodynamic configuration of high speed aircraft to solve the aerodynamic problems in high speed aircraft

4) Cultivate professional qualities of rigor, carefulness, and ability to express, coordinate, and communicate effectively

任务内容

1）薄翼型

2）薄翼型——层流翼型

3）薄翼型——超临界翼型

4）后掠翼

5）后掠翼的缺点

6）小展弦比机翼

7）小展弦比机翼的缺点

8）涡流发生器

9）涡流发生器的优点

 任务目标

1）理解高速飞机气动外形的特点

2）掌握高速飞机气动外形的工作原理

3）运用高速飞机气动外形的概念解决高速飞行中的空气动力学问题

4）培养严谨、细心的职业素养，以及有效表达、协调和沟通的能力

 Learning Guide

The high speed aircraft usually refers to high speed subsonic aircraft and supersonic aircraft. On high subsonic aircraft, local shock waves may occur due to excessive the flight speed. On supersonic aircraft, shock waves are definitely generated. These two types of aircraft are designed to address the shock waves caused by the increase in flight speed and the increase in drag caused by the shock waves. The solution is not only to have a high thrust jet engine, but also to have an aircraft shape that can adapt to the requirements of high speed flight.

课文

1. Thin Airfoil

1. 薄翼型

The airfoils of high speed aircraft wings are thin airfoils with relatively small thickness and backward maximum thickness point. The flight speed is high enough for high speed aircraft to obtain sufficient lift, thus high lift coefficient is not the primary issue. How to increase the critical Mach number and reduce the wave drag is more important.

With relatively small thickness airfoil, the airflow acceleration on the upper airfoil is effectively reduced, which greatly enhances the critical Mach number and the maximum flight speed of the aircraft. In addition, when entering transonic flight, the shock wave drag will increase with the relative thickness of the airfoil.

高速飞机机翼的翼型是相对厚度较小、最大厚度点位置靠后的薄翼型。由于飞行速度足以使高速飞机获得足够的升力，因此并不需要高的升力系数。如何提高临界马赫数和减小激波阻力更为重要。

对于相对厚度较小的翼型，上翼面的气流加速度有效降低，大大提高了飞机的临界马

赫数和最大飞行速度。当飞机进入跨声速飞行时，激波阻力将随着翼型相对厚度的增加而增大。

2. Thin Airfoil—Laminar Airfoil

2. 薄翼型——层流翼型

Laminar airfoil is widely used in high subsonic aircraft (Fig. 5-13). The leading edge radius of this airfoil is small, the position of the maximum thickness is backward, the airflow acceleration on the upper airfoil is slow, and the pressure distribution is flat. It plays an important role in improving the critical Mach number, so it is more suitable for high subsonic flight.

层流翼型广泛应用于高亚声速飞机（图 5-13）。这种翼型的前缘半径较小，最大厚度位置靠后，上翼面的气流加速较慢，压力分布平坦。它在提高临界马赫数方面起着重要作用，因此更适合于高亚声速飞行。

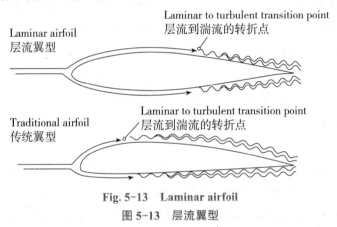

Fig. 5-13　Laminar airfoil

图 5-13　层流翼型

3. Thin Airfoil—Supercritical Airfoil

3. 薄翼型——超临界翼型

Supercritical airfoil can effectively increase the critical Mach number and has good aerodynamic characteristics in the transonic region (Fig. 5-14). Compared with the traditional airfoil, the upper airfoil of supercritical airfoil is flat and the rear part is slightly bent downward. Thus, the airflow acceleration on the upper surface is slow and the critical Mach number is enhanced.

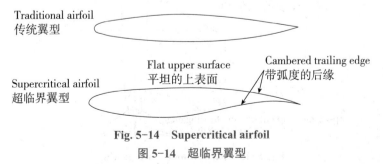

Fig. 5-14　Supercritical airfoil

图 5-14　超临界翼型

When the local supersonic region appears, the expansion and acceleration of supersonic airflow is relatively slow, which greatly reduces the local shock wave intensity. The location of the local shock wave is backward and the rear of the airfoil bends downward, which can alleviate the boundary layer separation induced by the shock wave and greatly reduce the drag of the transonic shock wave.

There is an reverse surface at the rear of the supercritical airfoil, which can increase the rear lift to make up for the lack of lift caused by the flat upper surface. Compared with laminar airfoil, its transonic aerodynamic characteristics are better.

超临界翼型可以有效地提高临界马赫数，在跨声速区域具有良好的气动特性（图 5-14）。与传统翼型相比，超临界翼型的上翼面较平坦，后部略向下弯曲。因此，上翼面的气流加速较慢，临界马赫数较大。

当局部超声速区域出现时，超声速气流的膨胀和加速相对较慢，这大大降低了局部激波强度。局部激波的位置靠后，翼型后部向下弯曲，这些都可以减轻激波引起的附面层分离，大大减少跨声速激波阻力。

超临界翼型下表面后部有一个向里凹进的反曲面，可增加后部升力，以弥补上表面平坦造成的升力不足。与层流翼型相比，其跨声速气动特性更好。

4. Swept Wing

4. 后掠翼

The critical Mach number of the aircraft can be increased and the wave drag can be reduced by using the swept wing configuration. For a swept wing, since the direction of the airflow speed is not perpendicular to the leading edge of the wing, it can be decomposed into the speed perpendicular and parallel to the leading edge of the wing (Fig. 5-15).

使用后掠翼构型，可以增加飞机的临界马赫数，减少激波阻力。对于后掠翼，由于空气速度的方向不垂直于机翼前缘，可以将其分解为垂直和平行于机翼前缘的速度（图 5-15）。

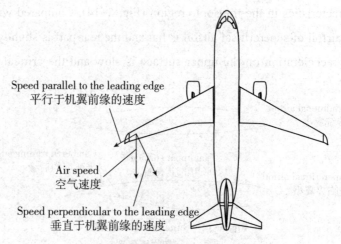

Speed parallel to the leading edge
平行于机翼前缘的速度

Air speed
空气速度

Speed perpendicular to the leading edge
垂直于机翼前缘的速度

Fig. 5-15　Airflow on the swept wing
图 5-15　后掠翼上的气流

The airflow parallel to the leading edge of the wing has no contribution to lift, only the airflow perpendicular to the leading edge of the wing generates lift. In this way, the airflow utilized by the airfoil is only part of the total. If this part of the airflow is accelerated to the local sound speed, the total airflow speed can be higher than that of a straight wing. Therefore, the swept wing can increase the critical Mach number of the aircraft. The greater the sweep angle, the more effective of increasing the critical Mach number.

The swept wing can also improve the transonic aerodynamic characteristics of the wing and reduce the wave drag. The critical Mach number of the swept wing is significantly improved, compared with the straight wing.

与机翼前缘平行的气流对升力没有贡献，只有垂直于机翼前缘的气流产生升力。这样，翼型所利用的气流仅为总气流的一部分。如果这部分气流被加速到局部声速，则总气流速度就可以高于平直翼。因此，后掠翼可以提高飞机的临界马赫数。后掠角越大，提高临界马赫数的效果越明显。

后掠翼还可以改善机翼的跨声速气动性能，减少激波阻力。与平直翼相比，后掠翼的临界马赫数显著提高。

5. Disadvantages of Swept Wing

5. 后掠翼的缺点

(1) The lift of swept wing in low speed flight are insufficient. Compared with the straight wing, the airflow used by the swept wing to generate lift is reduced, and the lift coefficient are also reduced. When flying at a low speed, the aircraft cannot generate enough lift. High-lift device and long runway length are necessary during low speed flight.

(2) The stall characteristics of swept wing are not good. When the airflow flows through the swept wing, due to the airflow parallel to the leading edge of the wing is flowing along the wing span direction, the boundary layer separation first occurs at the wingtip, which may cause the aircraft stall at high angle of attack and deteriorate the control efficiency of the aileron.

(3) The force distribution of the swept wing is not good. The sweep angle used by high subsonic civilian transport aircraft is not too large, which is generally about 30° and is mainly used to improve the critical Mach number.

（1）低速飞行时后掠翼的升力不足。与平直翼相比，后掠翼用于产生升力的气流减少，升力系数也降低了。以低速飞行时，飞机不能产生足够的升力。飞机在低速飞行期间需要增升装置和较长的跑道长度。

（2）后掠翼的失速特性不好。当气流流经后掠翼时，平行于机翼前缘的气流沿翼展方向流动，造成附面层分离首先在翼梢处发生，这可能导致飞机大迎角失速，并降低副翼的操纵效率。

（3）后掠翼的应力分布不好。高亚声速民用运输机使用的后掠角不会太大，通常约为30°，主要用于提高临界马赫数。

6. Low Aspect Ratio Wing

6. 小展弦比机翼

In order to reduce the induced drag, subsonic aircraft usually use high aspect ratio wings. For aircraft flying at transonic and supersonic speeds, the aspect ratio is greatly reduced and becomes a low aspect ratio wing.

The low aspect ratio wing (Fig. 5-16) is with long chord length and short wingspan. If the chord length is longer, the relative thickness of the airfoil can be reduced without changing the maximum thickness of the airfoil, so that the airflow accelerates slowly on the airfoil surface, which improves the critical Mach number. In addition, the short wingspan reduces the shock wave generated along the front and trailing edges of the wing, reduces the shock wave length and thus reduces the wave drag.

为了减少诱导阻力，亚声速飞机通常使用大展弦比机翼。对于以跨声速和超声速飞行的飞机，展弦比大大减小，成为小展弦比机翼。

小展弦比机翼（图 5-16）具有长弦长和短翼展。如果弦长较长，则可以在不改变翼型最大厚度的情况下减小翼型的相对厚度，从而使气流在翼型表面缓慢加速，提高临界马赫数。此外，短翼展减少了沿机翼前后缘产生的激波，减少了激波长度，从而减少了激波阻力。

Fig. 5-16　Low aspect ratio wing
图 5-16　小展弦比机翼

7. Disadvantages of Low Aspect Ratio Wing

7. 小展弦比机翼的缺点

The disadvantages of low aspect ratio is that when flying at low speeds, the induced drag can result in poor takeoff and landing performance.

小展弦比机翼的缺点是当以低速飞行时，诱导阻力会导致起飞和着陆性能变差。

8. Vortex Generator

8. 涡流发生器

Vortex generator is a device that uses vortex to bring energy into the boundary layer from external airflow, accelerates the airflow in the boundary layer and prevents air separation (Fig. 5-17).

涡流发生器是一种利用涡流将外部气流的能量引入附面层，加快附面层内的气流流动，并防止气流分离的装置（图 5-17）。

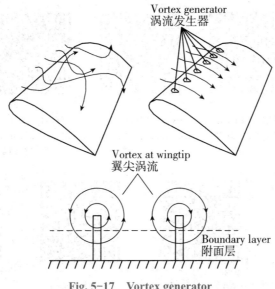

Fig. 5-17　Vortex generator
图 5-17　涡流发生器

The vortex generator is a small wing with low aspect ratio, which is vertically installed on the aerodynamic surface where they work. It can be arranged in pairs or in one direction. These small wings should have a certain angle of attack with the incoming airflow. When the air flows through the small wings at a certain angle of attack, it accelerates on one side and decelerates on the other side, resulting in a pressure difference on both sides of the small wing, and a strong wingtip vortex is generated at the end of the small wing. These vortices bring energy from the external airflow into the boundary layer, accelerate it and effectively inhibit the separation of it.

涡流发生器是一种低展弦比的小翼段，垂直安装在它们工作的气动表面上，可以成对布置或沿一个方向布置。这些小翼段都应与来流形成一定的迎角。当气流以一定的迎角流过小翼段时，在一侧加速，在另一侧减速，导致小翼段两侧产生压差，在小翼段末端产生强烈的翼尖涡流。这些涡流将外部气流的能量带入附面层，加速附面层气流流动并有效抑制附面层分离。

9. Advantages of Vortex Generators
9. 涡流发生器的优点

The vortex generator can be installed on the aerodynamic surface of low speed aircraft to prevent boundary layer separation. It can also be used in high subsonic and transonic aircraft to prevent or reduce boundary layer separation induced by shock, and improve the transonic aerodynamic characteristics of aircraft.

涡流发生器可以安装在低速飞机的气动表面，以防止附面层分离。它还可用于高亚声速和跨声速飞机，以防止或减少激波引起的附面层分离，并改善飞机的跨声速气动特性。

| Aerodynamic Configuration of High Speed (1) | Aerodynamic Configuration of High Speed (2) | Aerodynamic Configuration of High Speed (3) | Aerodynamic Configuration of High Speed (4) | Aerodynamic Configuration of High Speed (5) |

| Aerodynamic Configuration of High Speed (6) | Aerodynamic Configuration of High Speed (7) | Aerodynamic Configuration of High Speed (8) | Aerodynamic Configuration of High Speed (9) |

 New Words

sufficient	[sə'fɪʃnt]	*adj.*	足够的，充足的
primary	['praɪmeri]	*n.*	初选，第一
		adj.	初级的，主要的，最重要的
effectively	[ɪ'fektɪvli]	*adv.*	有效地，实际上，事实上
laminar airfoil	['læmɪnər 'erfɔɪl]		层流翼型
radius	['reɪdiəs]	*n.*	半径
pressure distribution	['preʃər ˌdɪstrɪ'bjuːʃn]		压力分布
supercritical	[ˌsupər'krɪtɪkəl]	*adj.*	超临界的
bend	[bend]	*n.*	拐弯，弯曲
		v.	弯曲，倾斜

appear	[əˈpɪr]	v.	出现，呈现
alleviate	[əˈliːvieɪt]	v.	减轻，缓和
induce	[ɪnˈduːs]	v.	诱导，引起
reverse	[rɪˈvɜːrs]	n.	相反的情况，后面，背面
		v.	颠倒，使相反
		adj.	相反的，反面的
lack	[læk]	n.	缺乏，匮乏
		v.	没有，缺乏，不足
characteristic	[ˌkærəktəˈrɪstɪk]	n.	特征，特点
		adj.	特有的，典型的
swept wing	[swept wɪŋ]		后掠翼
configuration	[kənˌfɪgjəˈreɪʃn]	n.	配置，结构，构造，布局
perpendicular	[ˌpɜːrpənˈdɪkjələr]	n.	垂直线
		adj.	垂直的，成直角的
contribution	[ˌkɒntrɪˈbjuːʃn]	n.	贡献，促成
utilize	[ˈjuːtəlaɪz]	v.	使用，利用
straight	[streɪt]	n.	直道
		adv.	直接，笔直地
		adj.	直的，准的
significantly	[sɪgˈnɪfɪkəntli]	adv.	显著地，明显地
disadvantage	[ˌdɪsədˈvæntɪdʒ]	n.	缺点，不利因素
insufficient	[ˌɪnsəˈfɪʃnt]	adj.	不充分的，不足的
runway	[ˈrʌnweɪ]	n.	飞机跑道
span	[spæn]	n.	跨度，范围
		v.	跨越，横跨
deteriorate	[dɪˈtɪriəreɪt]	v.	变坏，退化
aileron	[ˈeɪlərɒn]	n.	副翼
stress	[stres]	n.	应力，压力
edge	[edʒ]	n.	边，边缘
vortex generator	[ˈvɔːrteks ˈdʒenəreɪtər]		涡流发生器
arrange	[əˈreɪndʒ]	v.	安排，排列
pairs	[perz]	n.	一双，一对
inhibit	[ɪnˈhɪbɪt]	v.	阻止，阻碍
install	[ɪnˈstɔːl]	v.	安装，设置

The following questions are for you to answer to assess the learning outcomes.

(1) Briefly describe the characteristics of thin airfoil.

(2) Summarize which airfoil can be used for aircraft with different flight speeds.

(3) Briefly introduce the advantages of swept wing.

(4) Briefly describe the shortcomings of swept wing.

(5) Briefly introduce the characteristics of low aspect ratio wing.

(6) Briefly introduce the function of vortex generator.

 Extended Reading

Vortex Generators: Preventing Stalls at High and Low Speeds

You might have seen vortex generators—those little fins that protrude from the front of an airliner like the Boeing 737–800, or a short takeoff and landing (STOL) aircraft.

These little fins are amazing. They create vortices just like your wingtips do. How does that help? The vortices pull high energy air into the boundary layer, which delays a stall. They're an integral part of many aircraft to lower stall speed—but did you know that they're also used on transonic aircraft to keep control surfaces effective at high speeds?

1. Vortex Generators Create Mini Wingtip Vortices to Energize the Boundary Layer

Boundary layer is a layer of air right above the surface of the aircraft where skin friction slows down and removes energy from the airflow.

As air flows across the wing, the pressure decreases until it reaches the center of lift—about 25% down the wing's chord. Then, pressure starts to increase again, so the air moves from an area of low pressure to high pressure—this is called an "adverse pressure gradient". As the airflow moves towards high pressure, it loses energy. Eventually, when it runs out of energy, the airflow separates from the wing.

The air above the boundary layer isn't affected by skin friction, so it has more energy than the air in the boundary layer. If you could pull some of that free–stream air into the boundary layer, you could add energy and delay the boundary layer's separation. That's where the vortex generators come in.

Vortex generators act like tiny wings and create mini wingtip vortices, which spiral through the boundary layer and free–stream airflow. These vortices mix the high–energy free–stream air into the low–energy boundary layer, allowing the airflow in the boundary layer to withstand the adverse pressure gradient longer. Your wing can now operate at a higher angle of attack before airflow separation causes a stall.

On short takeoff and landing aircraft, you'll often see vortex generators along the leading edge of the wing. On airliners, you may see them in front of the flaps, where large adverse pressure gradients develop. In both cases, the vortex generators help keep the airflow attached at higher angles of attack, delaying a stall.

2. Vortex Generators Can also Delay a High Speed Stall

When airflow across an airfoil reaches transonic or supersonic speeds, a shock wave forms. Eventually, these shock waves will form at the leading edge of the airfoil, plus at the trailing edge and at any control surface hinge points.

As air moves across the shock wave, it suddenly loses energy. In fact, the energy loss may be so great that the airflow separates from the airfoil behind the shock wave—just like it does in a low speed stall. If an aileron or elevator lies behind the shock wave, the separated airflow makes the control surface ineffective, and it may make the aircraft impossible to control.

In this high speed situation, vortex generators can pull in high energy air from outside the boundary layer, mix it with air inside the boundary layer, and prevent separation. They can also disrupt the shock wave, reducing the amount of energy lost as air travels through the wave.

The horizontal stabilizer on a L-39 Albatros is a great example. A horizontal stabilizer is essentially an upside down wing that generates lift downward. Even though the L-39 is a subsonic aircraft, airflow moving over the tail can accelerate to transonic speeds, forming a shock wave. The vortex generators on the bottom of the stabilizer keep the airflow attached to the airfoil as it travels across the elevator, allowing you to maintain pitch control at high speeds.

飞行原理基础
Flight Fundamentals

Contents

学习内容

任务 1　飞行自由度
Task 1　Flight Freedom Degree

Contents

1) Center of gravity (CG)

2) Aircraft body coordinate system

3) Freedom degrees of CG

4) Freedom degrees of attitude

Learning Outcomes

1) Master the definition and position of the center of gravity of aircraft

2) Master the definition and direction of aircraft coordinate system

3) Master the definition and physical meaning of the six freedom degrees of aircraft

4) Solve the dynamic problems in flight by the freedom degrees of aircraft motion

5) Cultivate professional qualities of rigor, carefulness, and ability to express, coordinate, and communicate effectively

任务内容

1）重心（CG）

2）飞机机体坐标系

3）重心自由度

4）姿态自由度

任务目标

1）掌握飞机重心的定义和位置

2）掌握飞机坐标系的定义和方向

3）掌握飞机六个自由度的定义和物理意义

4）运用飞机运动的自由度解决飞行中的动力学问题

5）培养严谨、细心的职业素养，以及有效表达、协调和沟通的能力

Learning Guide

Freedom degree refers to the number of variables in physics that describe a physical state and independently to affect the results of the physical state. The degree of freedom of motion is the minimum number of coordinates required to determine the position of a system in space. The motion of a car has three degrees of freedom, while the motion of an airplane in the air has six degrees of freedom.

 课文

1. Center of Gravity
1. 重心

The gravity of the aircraft consists of the gravity of aircraft body, equipments, fuel, cargos and passenger loads, which is represented by W.

The point at which the aircraft gravity acts is called the center of gravity in Fig. 6-1.

飞机的重力包括机身、设备、燃料、货物和乘客的重力，用 W 表示。

飞机重力的作用点被称为重心，如图 6-1 所示。

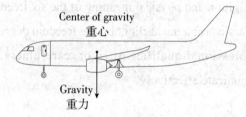

Fig. 6-1   The center of gravity
图 6-1   飞机的重心

Since the aircraft structure and onboard loads are basically symmetrical, the center of gravity is in the symmetry plane of the aircraft (Fig. 6-2).

由于飞机结构和机载载荷基本对称，所以飞机重心在机体对称平面内（图 6-2）。

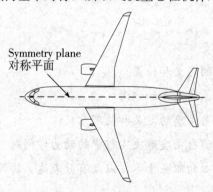

Fig. 6-2   The symmetry plane of the aircraft
图 6-2   机体对称平面

It can be assumed that all the mass of the aircraft is concentrated on its center of gravity, and the trajectory of the CG is used to represent the trajectory of the entire aircraft.

假设飞机的所有质量都集中在其重心上，可以用重心的运动轨迹表示整架飞机的运动轨迹。

2. Aircraft Body Coordinate System

2. 飞机机体坐标系

(1) Take the CG of the aircraft as the origin to construct the coordinate system.

(2) The origin of the coordinate system is at the CG of the aircraft.

(3) The longitudinal axis is along the fuselage of the aircraft and points to the aircraft nose.

(4) The lateral axis is perpendicular to the symmetry plane, from left to the right of the aircraft.

(5) The vertical axis is perpendicular to the longitudinal and lateral axis and points to the cabin.

The aircraft body coordinate system is shown in Fig. 6-3.

（1）以飞机重心为原点构建坐标系。

（2）坐标系原点位于飞机重心。

（3）纵轴平行于飞机机身，指向机头。

（4）横轴垂直于飞机对称面，从飞机的左侧到右侧。

（5）垂直轴垂直于横轴和纵轴，指向机舱上方。

飞机机体坐标系如图 6-3 所示。

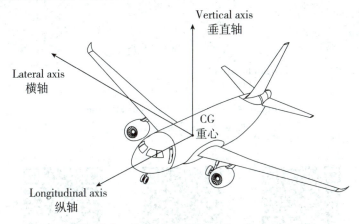

Fig. 6-3　Aircraft body coordinate system

图 6-3　飞机机体坐标系

3. Freedom degrees of CG

3. 重心自由度

There are three aircraft freedom degrees of CG.

(1) Translate along the longitudinal axis.

(2) Translate along the lateral axis.

(3) Translate along the vertical axis.

飞机重心自由度有三个。

（1）沿着纵轴的平移。

（2）沿着横轴的平移。

（3）沿着垂直轴的平移。

4. Freedom degrees of Attitude

4. 姿态自由度

There are also three aircraft freedom degrees of attitude (Fig. 6-4).

(1) Pitch around the lateral axis.

(2) Roll around the longitudinal axis.

(3) Yaw around the vertical axis.

飞机姿态自由度也有三个（图 6-4）。

（1）围绕横轴的俯仰。

（2）围绕纵轴的滚转。

（3）围绕垂直轴的偏航。

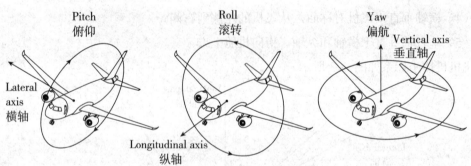

Fig. 6-4　Freedom degrees of attitude

图 6-4　飞机姿态自由度

Flight Freedom Degree (1)　　Flight Freedom Degree (2)　　Flight Freedom Degree (3)

160

New Words

gravity	['grævəti]	n.	重力，地球引力
mass	[mæs]	adj.	大批的，数量极多的，广泛的
concentrate	['kɒnsntreɪt]	v.	集中 (注意力)，聚精会神
trajectory	[trə'dʒektəri]	n.	轨迹，弹道
origin	['ɒrɪdʒɪn]	n.	起源，源头，起因
coordinate	[kəʊ'ɔːdɪneɪt]	n.	坐标
longitudinal	[ˌlɒŋgɪ'tjuːdɪnl]	adj.	纵向的，纵观的，经度的
axis	['æksɪs]	n.	轴 (旋转物体假想的中心线)
lateral	['lætərəl]	adj.	侧面的，横向的
vertical	['vɜːtɪkl]	adj.	竖的，垂直的，直立的
freedom	['friːdəm]	n.	自由，自由度
attitude	['ætɪtjuːd]	n.	态度，看法，姿势

 Q&A

The following questions are for you to answer to assess the learning outcomes.

(1) Besides the center of gravity, which special points of aircraft have you learned?

(2) Briefly describe how the coordinate axis of aircraft is constructed.

任务 2　飞行姿态
Task 2　Flight Attitude

 Contents

1) Attitude of the aircraft

2) Ground coordinate system

3) Flight attitude to the ground

4) Flight attitude to the relative airflow

Learning Outcomes

1) Master the definition of aircraft attitude relative to the ground

2) Master the definition of aircraft attitude relative to the incoming airflow

3) Solve the dynamic problems in flight by using the attitude of aircraft motion

4) Cultivate professional qualities of rigor, carefulness, and ability to express, coordinate, and communicate effectively

 任务内容

1）飞机的姿态
2）地面坐标系
3）相对地面的飞行姿态
4）相对来流的飞行姿态

 任务目标

1）掌握飞机相对于地面姿态的定义
2）掌握飞机相对于来流姿态的定义
3）运用飞机运动姿态解决飞行中的动力学问题
4）培养严谨、细心的职业素养，以及有效表达、协调和沟通的能力

Learning Guide

Flight attitude refers to the state of the three axes of an aircraft in the air relative to a reference line or plane, or a fixed coordinate system. Aircraft flying in the air is different from vehicles moving on the ground, as they have various flight postures. This refers to changes in the aircraft's tilt, bow, left tilt, right tilt, etc. The flight attitude determines the direction of an aircraft, affecting both the altitude and direction of flight. When flying at low speeds, the pilot can determine the aircraft's attitude based on the position of the horizon by observing the ground.

课文

1. Attitude of the Aircraft
1. 飞机的姿态

The attitude of the aircraft can be determined by the directional relationship between the body coordinate system and the ground coordinate system.

The attitude of the aircraft is represented by Euler angle (Fig. 6-5).

飞机的姿态可以通过机体坐标系和地面坐标系之间的方向关系来确定。

飞机的姿态用欧拉角表示（图 6-5）。

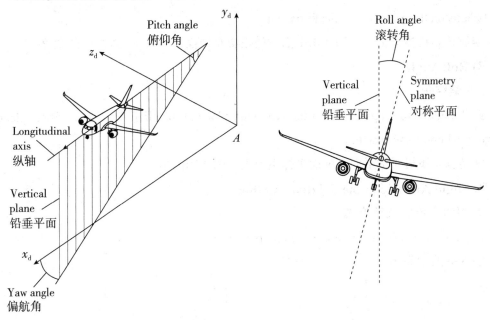

Fig. 6-5　Euler angle
图 6-5　欧拉角

2. Ground Coordinate System

2. 地面坐标系

The ground coordinate system is fixed on the earth's surface.

The ground coordinate system is $Ax_dy_dz_d$.

The origin is at a fixed point on the ground, Ay_d axis is perpendicular to the earth, Ax_d and Az_d are perpendicular to each other in the horizontal plane.

地面坐标系固定在地球表面。

地面坐标系为 $Ax_dy_dz_d$。

原点位于地面上的一个固定点。Ay_d 轴垂直于地面，Ax_d 和 Az_d 在水平面上相互垂直。

3. Flight Attitude to the Ground

3. 相对地面的飞行姿态

1) Pitch Angle

1）俯仰角

Pitch angle is the angle between the longitudinal axis of the aircraft and the horizontal plane.

俯仰角是机体坐标系纵轴与对称平面之间的角度。

2) Yaw Angle

2）偏航角

Yaw angle is the angle between the projection of longitudinal axis on horizontal plane and the Ax_d axis on the ground coordinate system.

偏航角是机体坐标系纵轴在水平面上的投影与地面坐标系 Ax_d 轴之间的角度。

3) Roll Angle

3）滚转角

Roll angle is the angle between the symmetry plane of the aircraft and the vertical plane on the ground coordinate system.

滚转角是飞机对称平面与地面坐标系铅垂平面之间的角度。

4. Flight Attitude to the Relative Airflow

4. 相对气流的飞行姿态

Fig. 6-6 shows flight altitude to the relative airflow.

图 6-6 所示为相对气流的飞行姿态。

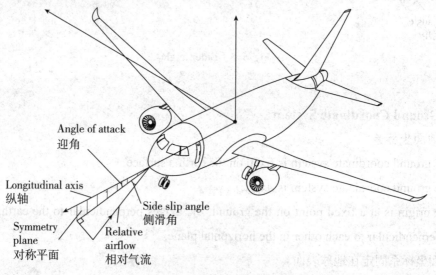

Fig. 6-6   Flight attitude to the relative airflow

图 6-6   飞机相对气流的飞行姿态

(1) Angle of attack is the angle between the relative airflow and the longitudinal axis of the aircraft in the symmetry plane of the aircraft.

(2) Side slip angle is the angle between the relative airflow and the aircraft symmetry plane.

(3) Roll and yaw of the aircraft will cause side slip, and the external disturbance will also cause side slip.

(4) In some cases, appropriate side slip angle is beneficial to flight, such as crosswind landing, asymmetric power flight, etc.

164

（1）迎角是飞机对称平面内来流与飞机纵轴之间的角度。

（2）侧滑角是来流与飞机对称面之间的角度。

（3）飞机的横滚和偏航会导致侧滑，外部干扰也会导致侧滑。

（4）在某些情况下，采用适当的侧滑角有利于飞行，如侧风着陆、不对称动力飞行，等等。

Flight Attitude (1)

Flight Attitude (1)

Flight Attitude (1)

 New Words

body coordinate system　　['bɒdi kəʊ'ɔːdɪneɪt 'sɪstəm] 机体坐标系

ground coordinate system　　[graʊnd kəʊ'ɔːdɪneɪt 'sɪstəm] 地面坐标系

fixed	[fɪkst]	*adj.* 固定的，不变的
horizontal	[ˌhɒrɪ'zɒntl]	*adj.* 水平的，与地面平行的，横的
plane	[pleɪn]	*n.* 飞机，平面
projection	[prə'dʒekʃn]	*n.* 投影
side slip	['saɪd ˌslɪp]	*n.* 横滑，侧滑
appropriate	[ə'prəʊpriət]	*adj.* 适当的，合适的
crosswind	['krɒswɪnd]	*n.* 侧风

 Q&A

The following questions are for you to answer to assess the learning outcomes.

(1) Describe the definition of side slip angle.

(2) Describe three attitudes of an aircraft around the coordinate axis.

任务 3 飞行的负载和平衡
Task 3 Flight Loads and Balance

 Contents

1) Loads and balance

2) Steady flight

 Learning Outcomes

1) Master the force of the aircraft in flight

2) Master the balance of the aircraft in flight

3) Solve the dynamic problems in flight by the force and balance in flight

4) Cultivate professional qualities of rigor, carefulness, and ability to express, coordinate, and communicate effectively

 任务内容

1）负载和平衡

2）定常飞行

 任务目标

1）掌握飞机飞行中的受力情况

2）掌握飞机飞行中的平衡情况

3）运用飞机飞行中的受力和平衡解决飞行中的动力学问题

4）培养严谨、细心的职业素养，以及有效表达、协调和沟通的能力

 Learning Guide

The state in which the sum of all external forces and moments acting on the aircraft during flight is equal to zero is called force balance, also known as the equilibrium state of the aircraft. The problem of force balance in aircraft can be summarized as longitudinal balance, lateral balance, and heading balance. It is the key to stable and safe flight of aircraft.

1. Loads and Balance

1. 负载和平衡

In flight, the external forces acting on the aircraft are gravity, aerodynamic forces, which contains lift, drag, lateral force, and thrust (Fig. 6–7).

To balance these external forces, some terms are required to be met as follows.

(1) Term 1: The resultant force is zero, namely

$$\sum X=0, \quad \sum Y=0, \quad \sum Z=0$$

(2) Term 2: The resultant moment is zero, namely

$$\sum M_X=0, \quad \sum M_Y=0, \quad \sum M_Z=0$$

Therefore, the external forces acting on the aircraft should meet six equilibrium equations when they reach the balance state.

There are six degrees of freedom for aircraft maneuvering in the air.

在飞行中，作用在飞机上的外力是重力、空气动力（包括升力、阻力、侧向力和推力）（图 6-7）。

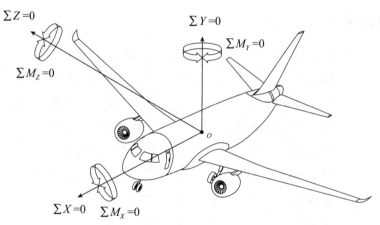

Fig. 6-7　Balance of the aircraft
图 6-7　飞机的平衡状态

为了平衡这些外力，需要满足以下条件。

（1）第 1 项：合力为零，即

$$\sum X=0, \quad \sum Y=0, \quad \sum Z=0$$

（2）第 2 项：合力矩为零，即

$$\sum M_X=0, \quad \sum M_Y=0, \quad \sum M_Z=0$$

因此，当飞机达到平衡状态时，作用在飞机上的外力应满足六个平衡方程。

飞机在空中运动有六个自由度。

2. Steady Flight

2. 定常飞行

Steady flight means that the aircraft is in a balanced flight state, its speed and flight direction remain unchanged.

Steady flight is the most frequent and important flight state of the aircraft.

Steady flight includes steady cruise flight, steady climb, steady descend, etc.

If the external forces acting on the aircraft cannot meet the balance requirements, the aircraft will make variable movement, the speed or direction of it would change, thereby changing the flight state.

定常飞行是指飞机处于平衡飞行状态，其速度和飞行方向保持不变。

定常飞行是飞机最频繁、最重要的飞行状态。

定常飞行包括稳速巡航飞行、稳速爬升、稳速下滑等。

如果作用在飞机上的外力不能满足平衡要求，飞机将进行变速运动，其速度或方向将发生变化，从而改变原来的飞行状态。

Flight Loads and Balance

 New Words

external	[ɪk'stɜːnl]	adj.	外部的，外面的
term	[tɜːm]	n.	条件，学期，术语，期限
resultant	[rɪ'zʌltənt]	n.	结果，合力，后果
equilibrium	[ˌiːkwɪ'lɪbriəm]	n.	平衡，均衡，均势
steady	['stedi]	adj.	稳步的，持续的
frequent	[fri'kwent]	adj.	频繁的，经常发生的
climb	[klaɪm]	v.	攀登，爬升
glide	[glaɪd]	v.	滑行，滑动
variable	['veəriəbl]	adj.	可变的，多变的

 Q&A

The following questions are for you to answer to assess the learning outcomes.

(1) What requirements should the external force on the aircraft meet when the aircraft is in equilibrium?

(2) Describe the definition of steady flight.

任务 4 载荷系数
Task 4 Load Factors

 Contents

    1) Definition of the load factor

    2) Decomposition of load factor

    3) Meaning of the load factor

    4) Application of the load factor

Learning Outcomes

    1) Master the definition and physical meaning of the load factor

    2) Analyze the overload of the aircraft under different flight conditions

    3) Solve the dynamic problems in flight by the overload conditions

    4) Cultivate professional qualities of rigor, carefulness, and ability to express, coordinate, and communicate effectively

 任务内容

    1）载荷系数的定义

    2）载荷系数的分解

    3）载荷系数的含义

    4）载荷系数的应用

任务目标

    1）掌握载荷系数的定义和物理意义

    2）分析飞机在不同飞行状态下的过载情况

    3）运用飞机飞行中的过载情况解决飞行中的动力学问题

    4）培养严谨、细心的职业素养，以及有效表达、协调和沟通的能力

 Learning Guide

    Overload can be understood as the maximum acceleration that the aircraft's body structure can withstand during high speed flight, especially during various high speed

and intense maneuvers. Usually, the larger the overload of an aircraft, the more robust the aircraft structure needs to be and the more sensitive the control needs to be. Comparing the turning radius on an aircraft is meaningless because once the speed increases, the turning radius also increases accordingly. Comparing the angular velocity is also meaningless because it is also related to speed.

 课文

1. Definition of the Load Factor
1. 载荷系数的定义

The load factor is a dimensionless parameter used to explain the forces on the aircraft under various flight conditions.

The load factor is defined by the ratio of the external load component (excluding weight) acting on the aircraft to the aircraft gravity.

载荷系数是一个无量纲参数，用于解释飞机在各种飞行条件下所受的力。

载荷系数定义为作用在飞机上的外载荷分量（不包括飞机重力）与飞机重力之比。

2. Decomposition of the Load Factor
2. 载荷系数的分解

The load factor can be decomposed into three directions, which are the same as the direction of the coordinate axis of the aircraft body coordinate system.

According to the aircraft body coordinate system, the load factor are n_X, n_Y, n_Z, respectively.

n_Y is defined as follows:

$$n_Y = \frac{L}{W}$$

Where L is lift, and W is aircraft gravity.

载荷系数可以分解为三个方向，与飞机机体坐标系的坐标轴方向相同。

根据飞机机体坐标系，载荷系数分别为 n_X, n_Y, n_Z。

n_Y 定义如下：

$$n_Y = \frac{L}{W}$$

式中，L 为升力；W 为飞机重力。

3. Meaning of the Load Factor
3. 载荷系数的含义

The value of the load factor in one direction indicates how many times the external load in this direction (excluding aircraft gravity) is to the aircraft gravity.

The positive and negative load factor indicate the direction of the external loads.

某个方向载荷系数的值表示该方向的外载荷（不包括飞机重力）是飞机重力的多少倍。

载荷系数的正和负表示外载荷的方向。

4. Application of the Load Factor

4. 载荷系数的应用

Generally speaking, "aircraft overload" is generally described by the load factor n_Y, since among the three load factors, the one that varies greatly in flight and has the greatest impact on the structure is along the vertical axis of aircraft body coordinate system.

(1) In level flight, $L=W$, $n_Y=1$.

(2) When the aircraft pulls up in a dive, the lift of the aircraft not only overcomes its own gravity, but also provides centripetal force for the aircraft to move along a curvilinear trajectory. In this case, the lift can reach several times of the aircraft gravity.

$$n_Y = \frac{L}{W}, \ n_Y \geqslant 1$$

(3) When the aircraft dives, the lift is smaller than the aircraft gravity, sometimes it can be negative.

$$n_Y = \frac{L}{W}, \ n_Y < 1$$

(4) Maneuvering overload refers to the overload caused by the change of lift when the pilot controls the aircraft during the maneuvering flight, such as entering into a dive, pulling up in a dive, and turning horizontally, etc.

(5) Gust overload refers to the overload caused by gust, especially vertical gust, which changes the flight speed and angle of attack of the aircraft in flight, thereby causing the change of the aircraft lift.

(6) If a large vertical upward gust is encountered, a relatively large positive overload will occur. If there is a relatively large vertical downward gust, it will produce a large negative overload.

"飞机过载"一般用载荷系数 n_Y 描述，在三个载荷系数中，在飞行中变化最大且对结构影响最大的载荷系数是沿飞机机体坐标系垂直轴方向的载荷因子。

（1）在平飞中，$L=W$，$n_Y=1$。

（2）当飞机做俯冲拉起时，飞机的升力不仅克服了其重力，还为飞机提供了沿曲线轨迹移动的向心力。在这种情况下，升力可以达到飞机重力的几倍。

$$n_Y = \frac{L}{W}, \ n_Y \geqslant 1$$

（3）当飞机俯冲时，升力小于飞机的重力，有时升力为负值。

$$n_Y = \frac{L}{W}, \ n_Y < 1$$

（4）机动过载是指当飞行员操纵飞机在机动飞行过程中，由于升力变化而引起的过载，如进入俯冲、俯冲拉起、水平转弯等。

（5）突风过载是指由于突然的风，特别是垂直突风引起的过载，它改变了飞行中飞机的飞行速度和迎角，进而引起飞机升力的变化。

（6）如果遇到较大的垂直向上的突风，将发生相对较大的正过载。如果有相对较大的垂直向下的突风，将产生较大的负过载。

Load Factors (1)　　　Load Factors (2)

 New Words

load factor	['ləʊd 'fæktə(r)]		载荷系数
ratio	['reɪʃiəʊ]	*n.*	比率，比例
decompose	[diːkəm'pəʊz]	*v.*	（使）分解，腐烂
respectively	[rɪ'spektɪvli]	*adv.*	分别地，分别
times	[taɪmz]	*n.*	（用于比较）倍
dive	[daɪv]	*v.*	俯冲，突然下降
pull up	['pʊl ʌp]		拉起
overcome	[ˌəʊvə'kʌm]	*v.*	克服，解决，战胜
centripetal	[ˌsentrɪ'piːtl]	*adj.*	向心的
maneuver	[mə'nuːvə]	*n.*	机动，演习，调动
gust	[gʌst]	*n.*	突风
encounter	[ɪn'kaʊntə(r)]	*v.*	遭遇，遇到

 Q&A

The following questions are for you to answer to assess the learning outcomes.

(1) Describe the definition of load factor.

(2) What is the load factor when the aircraft is in level flight?

(3) What is the load factor when the aircraft dives?

(4) What is the load factor when the aircraft pulls up in a dive?

 Extended Reading

Aircraft Load Factor

This is an aircraft load factor graph (Fig. 6-8), demonstrating how *G* loading or *G*-forces occur on the aircraft, and in turn, when you making level turns (level turns being holding a constant altitude and not climbing or descending. As we talk about being level in this topic, what

172

we mean is maintaining the same altitude). As you increase the bank angle of the airplane, the lift being generated by the wing is no longer just pushing straight down to keep the airplane in the air, this lift is now being directed at an angle which is ultimately what makes your airplane turn. The horizontal component of lift is what makes the aircraft turn.

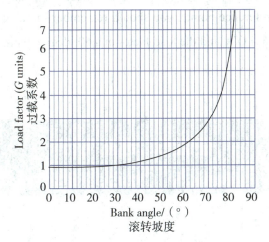

Fig. 6-8　Load factor with bank angle
图 6-8　过载系数与滚转坡度的关系

We can break the total lift being generated by the wing into two separate "vectors" or forces; the horizontal component, and the vertical component. The vertical component is what keeps the airplane flying level, and as a result, the vertical component must remain the same to keep the airplane flying level regardless of turning or flying straight. Now, to keep this vertical component the same when we are directing or lift off to the side by banking the airplane, we are going to have to increase our total lift on the wing, which in turn keeps the vertical lift vector constant, and increases the horizontal component of lift (making the airplane turn). This increase in the total lift (total lift in normal straight and level flight is about $1G$ force) is felt by the pilot and occupants of the aircraft as increased G-forces (Fig. 6-9).

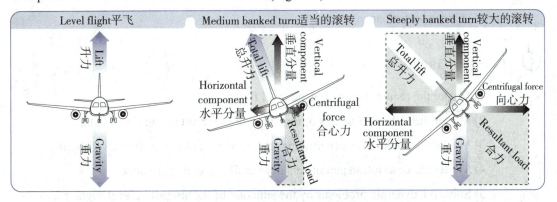

Fig. 6-9　Force with bank angle
图 6-9　不同滚转坡度下的飞机受力

Ultimately, the more the wing tilts (banks), the more lift it must generate to keep the airplane in level flight. Obviously once the airplane banks to 90 ° the amount of lift required becomes infinite, since the lift is only being directed sideways and no amount of lift from the wing (or pulling back on the controls by the pilot) could keep the airplane level.

If you want a few reference points, you should remember the load factor or G-forces in a 45-degree bank turn are 1.4 G's, and the load factor in a 60-degree bank turn is 2 G's.

任务 5　巡航飞行
Task 5　Cruise Flight

 Contents

    1) Required speed for level flight

    2) Factors that affect the speed required for level flight

    3) Required thrust for level flight

    4) Factors affecting the required thrust for level flight

    5) Required power for level flight

    6) Maximum level flight speed

    7) Factors affecting the maximum level flight speed

    8) Minimum level flight speed

    9) Factors affecting the minimum level flight speed

    10) Level flight speed range

    11) Cruise performance

 Learning Outcomes

    1) Master the requirements of plane level flight

    2) Master the influence factors of thrust required for aircraft in level flight

    3) Master the method for determining the speed range of aircraft in level flight

    4) Master the description parameters of aircraft cruise performance

    5) Solve the dynamic problems by the situation of the aircraft in level flight

    6) Cultivate professional qualities of rigor, carefulness, and ability to express,

coordinate, and communicate effectively

任务内容

1）平飞所需速度

2）平飞所需速度的影响因素

3）平飞所需推力

4）平飞所需推力的影响因素

5）平飞所需功率

6）最大平飞速度

7）最大平飞速度的影响因素

8）最小平飞速度

9）最小平飞速度的影响因素

10）平飞速度范围

11）巡航性能

任务目标

1）掌握飞机平飞的要求

2）掌握飞机平飞所需推力的影响因素

3）掌握飞机平飞速度范围的确定方法

4）掌握飞机巡航性能的描述参数

5）运用飞机平飞的情况解决飞行中的动力学问题

6）培养严谨、细心的职业素养，以及有效表达、协调和沟通的能力

Learning Guide

Cruise flight refers to the economical flight state chosen by an aircraft for long-distance and long-term flight missions. The cruising speed of an aircraft is related to many factors, such as flight distance, required time, load requirements, flight safety, engine durability, and economy. Due to the good economy of cruise flight, it is commonly used for long-distance flight missions, such as air transportation, observation, patrol, escort, and transition. The main ways to improve the aircraft's cruise performance are adopting aerodynamic layout with high lift drag ratio, improving engine performance, reducing fuel consumption, carrying auxiliary fuel tank or using aerial refueling, etc.

175

课文

1. Required Speed for Level Flight

1. 平飞所需速度

The speed required for level flight refers to the flight speed required to obtain the lift for level flight, when the aircraft is cruising at a steady speed at a certain altitude.

平飞所需速度是指飞机在某一高度以稳定速度巡航时，获得平飞升力所需的飞行速度。

2. Factors That Affect the Speed Required for Level Flight

2. 影响平飞所需速度的因素

The factors that affect the speed required for level flight are aircraft mass, wing area, air density and lift coefficient.

The heavier the aircraft mass, the higher the speed required for level flight.

When the aircraft is cruising, the aircraft mass, wing area, air density remain unchanged, the speed required for the aircraft in level flight is only related to the lift coefficient.

Since the lift coefficient varies with the angle of attack, the cruise speed of the aircraft mainly varies with the angle of attack at a certain cruising altitude. Reducing the angle of attack can increase the speed required for level flight. Increasing the angle of attack can reduce the speed required for level flight.

影响平飞所需速度的因素有飞机质量、机翼面积、空气密度和升力系数。

飞机质量越大，平飞所需的速度越高。

当飞机巡航时，飞机质量、机翼面积和空气密度不变，飞机平飞所需的速度仅与升力系数有关。

由于升力系数随迎角变化，因此在一定巡航高度下，飞机的巡航速度主要随迎角而变化。减小迎角可以增大平飞所需的速度，增大迎角可以减小平飞所需的速度。

3. Required Thrust for Level Flight

3. 平飞所需推力

When the aircraft flies with a certain speed at a certain altitude in level flight, it must fly at a certain angle of attack in order to balance the lift and gravity (Fig. 6-10). At that altitude, speed and angle of attack, the drag generated by level flight must be overcome by the thrust of the engine, which is called the required thrust for level flight.

当飞机在某一高度以一定速度平飞时，它必须以一定的迎角飞行，以平衡升力和重力（图6-10）。在这一高度、速度和迎角下，平飞产生的阻力必须由发动机的推力克服。这种推力被称为平飞所需推力。

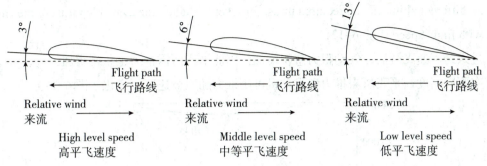

Fig. 6-10   Level flight speed and the corresponding angle of attack

图 6-10   飞机平飞速度和对应的迎角

4. Factors Affecting the Required Thrust for Level Flight

4. 平飞所需推力的影响因素

Since the required thrust for steady level flight is obtained according to the condition that the thrust is equal to the drag, the required thrust at different altitudes can be obtained from different flight altitudes.

Within a certain speed range, with the increase of level flight speed, the thrust required for level flight first decreases and then increases (Fig. 6-11).

由于定常平飞所需的推力是根据推力和阻力相等条件获得的，因此可以从不同的飞行高度获得不同高度所需推力。

在一定的速度范围内，随着平飞速度的增加，平飞所需推力先减小后增大（图6-11）。

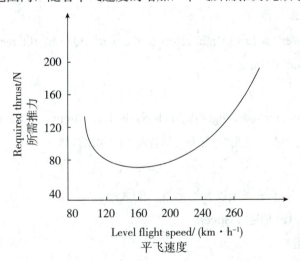

Fig. 6-11   The required thrust for level flight

图 6-11   平飞所需推力

After entering the transonic flight, the aerodynamic drag increases sharply, so the required thrust for level flight also increases sharply.

At different altitudes, the required thrust for level flight and the available thrust of the engine vary with flight speed (Fig. 6-12).

进入跨声速飞行后，气动阻力急剧增加，因此平飞所需推力也急剧增加。

在不同高度，平飞所需推力和发动机可用推力随飞行速度而变化（图6-12）。

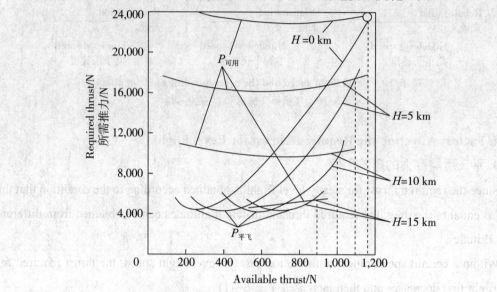

Fig. 6-12   Required thrust and the available thrust of the engine in level flight

图 6-12   平飞所需推力和发动机可用推力

5. Required Power for Level Flight

5. 平飞所需功率

The required power for level flight refers to the work done by the required thrust for level flight in a certain period of time.

$$P = N \cdot v$$

The required power for level flight depends on the level flight speed and the required thrust.

平飞所需功率是指在一定时间内平飞所需推力所做的功。

$$P = N \cdot v$$

平飞所需功率取决于平飞速度和所需推力的大小。

6. Maximum Level Flight Speed

6. 最大平飞速度

The maximum level flight speed generally refers to the maximum stable level flight speed that the aircraft can achieve in a straight and level flight with the engine in full throttle.

最大平飞速度通常是指飞机在发动机全速运转的情况下水平直线飞行时能够达到的最大稳定平飞速度。

7. Factors Affecting the Maximum Level Flight Speed

7. 最大平飞速度的影响因素

At different flight altitudes, the thrust required for level flight is different, and the available thrust of the engine under rated state is also different. Therefore, the maximum level flight speed of the aircraft is also different.

With the increase of altitude, the maximum level flight speed gradually decreases.

The maximum level flight speed of the aircraft is not only limited by the available thrust of the engine, but also related to the ability of the aircraft structure to bear aerodynamic loads.

When flying below cruising altitude, due to the limitation of aircraft structural strength, the flight speed of the aircraft cannot reach the maximum level flight speed allowed by the available thrust of the engine, that is, the level flight speed that the aircraft can reach is smaller than the maximum level flight speed.

在不同的飞行高度，平飞所需推力不同，发动机在额定状态下的可用推力也不同。因此，飞机的最大平飞速度也不同。

随着高度的增加，最大平飞速度逐渐减小。

飞机的最大平飞速度不仅受发动机可用推力的限制，还与飞机结构承受气动载荷的能力有关。

在巡航高度以下飞行时，由于飞机结构强度的限制，飞机的飞行速度无法达到发动机可用推力允许的最大平飞速度，即飞机能够达到的平飞速度小于最大平飞速度。

8. Minimum Level Flight Speed

8. 最小平飞速度

The minimum level flight speed is the lowest stable speed for an aircraft to maintain level flight.

最小平飞速度是飞机保持平飞的最小稳定速度。

9. Factors Affecting the Minimum Level Flight Speed

9. 影响最小平飞速度的因素

In order to obtain the lift required for level flight, the level flight speed reaches the minimum when the lift coefficient is maximum.

The minimum level flight speed is limited by the maximum lift coefficient.

When the lift coefficient is maximum, the angle of attack of the aircraft reaches the critical angle of attack, and the aircraft flight speed is the stall speed.

For flight safety, the aircraft cannot fly at the critical angle of attack, so the minimum level flight speed is higher than the stall speed.

The minimum level flight speed is not only limited by the maximum lift coefficient, but also related to the available thrust of the engine.

When the flight altitude gradually increases, at the critical angle of attack, the increased drag may exceed the available thrust of the engine and the minimum level flight speed may increase. At this point, the minimum level flight speed of the aircraft is limited by the available thrust of the engine.

为了获得平飞所需的升力，当升力系数最大时，飞机的平飞速度达到最小。

最小平飞速度受最大升力系数限制。

当升力系数最大时，飞机的迎角达到临界迎角，飞机的飞行速度为失速速度。

为了飞行安全，飞机不能在临界迎角飞行，因此最小平飞速度高于失速速度。

最小平飞速度不仅受最大升力系数的限制，而且与发动机的可用推力有关。

当飞机以临界迎角飞行，飞行高度逐渐增加时，增加的阻力可能超过发动机的可用推力，使最小平飞速度增加。此时，飞机最小平飞速度受到发动机可用推力的限制。

10. Level Flight Speed Range

10. 平飞速度范围

When the aircraft flies at a certain altitude, the range from the minimum level flight speed to the maximum level flight speed is called the level flight speed range of the aircraft at that flight altitude.

The minimum level flight speed and the maximum level flight speed of the aircraft vary with the flight altitude, so the level flight speed range also varies with the flight altitude.

The larger the level flight speed range of the aircraft, the better the level flight performance of the aircraft.

当飞机在某一高度飞行时，从最小平飞速度到最大平飞速度的范围称为该飞行高度的飞机平飞速度范围。

飞机的最小平飞速度和最大平飞速度随飞行高度的变化而变化，因此平飞速度范围也随飞行高度变化。

飞机平飞速度范围越大，飞机的平飞性能越好。

11. Cruise Performance

11. 巡航性能

The cruise performance of an aircraft is mainly cruise speed, flight range and flight time.

Cruise speed refers to the flight speed with the minimum fuel consumption per kilometer, that is, the flight speed corresponding to the maximum flight range.

Flight range refers to the horizontal distance that an aircraft can continuously fly without wind and refueling.

Flight time refers to the time that an aircraft can continuously fly when the available fuel is exhausted.

飞机的巡航性能主要是巡航速度、航程、航时。

巡航速度是指每公里燃油消耗量最小的飞行速度，即与最大航程相对应的飞行速度。

航程是指飞机在无风、不加油情况下能够连续飞行的水平距离。

航时是指飞机在可用燃油耗尽时能够持续飞行的时间。

Cruise Flight (1)

Cruise Flight (2)

Cruise Flight (3)

Cruise Flight (4)

Cruise Flight (5)

Cruise Flight (6)

Cruise Flight (7)

Cruise Flight (8)

 New Words

range	[reɪndʒ]	n. 范围，一系列
sharply	[ˈʃɒpli]	adv. 急剧地，严厉地，猛烈地
available	[əˈveɪləbl]	adj. 可用的，可获得的，可购得的
throttle	[ˈθrɒtl]	n. 油门，节流阀，节流杆
gradually	[ˈɡrædʒuəli]	adv. 逐步地，逐渐地，渐进地
bear	[beə(r)]	v. 承受，忍受
strength	[streŋθ]	n. 力量，强度
stall	[stɔːl]	n. 失速，停止
cruise speed	[kruːz spiːd]	巡航速度
consumption	[kənˈsʌmpʃn]	n. 消耗，消耗量
kilometer	[ˈkɪləʊˌmiːtə]	n. 千米
flight range	[flaɪt reɪndʒ]	航程，飞行距离
refueling	[ˌriːˈfjuːəlɪŋ]	v. 加油
flight time	[flaɪt taɪm]	航时，飞行时间

Q&A

The following questions are for you to answer to assess the learning outcomes.

(1) Describe the definition of the required speed for level flight.

(2) What are the factors that affect the required speed for level flight?

(3) Describe the definition of the required thrust for level flight.

(4) Describe the definition of the minimum level flight speed.

(5) Describe the definition of the cruise speed.

 Extended Reading

Managing Your Cruise: Understanding Speeds

Speed in cruise is often driven by performances and fuel burn considerations; however, air traffic or weather considerations sometimes intervene and require modifications to the optimum cruise profile. Whatever the flight crew's decisions to best optimize their flight, one needs to be constantly aware of the applicable limits and maneuvering speeds. To safely manage the cruise phase within the aircraft certified flight envelope, some characteristic speeds are useful references for flight crews to monitor the aircraft's actual speed. What speeds exactly should be monitored? What do these speeds mean and what happens if they are ignored?

Many speeds are used to certify and fly an aircraft operationally. For every flight, the applicable characteristic speeds are computed automatically by the aircraft autoflight systems (AFS)—flight management system (FMS), flight guidance (FG) and flight envelope (FE) and displayed on the PFD airspeed scale. They are extremely useful as target maneuvering and limit reference speeds to safely guide the pilots navigation decisions through the cruise phase (Fig. 6–13).

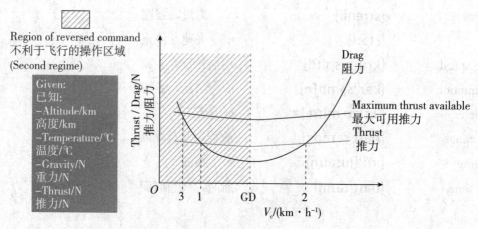

Fig. 6–13 Thrust curves and speed range
图 6–13 推力曲线和速度范围

GD speed is computed by the auto flight systems and is based on the aircraft mass [thanks to the zero fuel mass (ZFW) inserted in the FMS during flight preparation]. The GD formula has been set up so that the resulting airspeed provides the best lift-to-drag ratio for a given altitude, Mach number and aircraft weight, in clean configuration with one engine out.

In cruise:

Above GD, the drag and thrust required to maintain speed increase with the speed.

Below GD, the drag and thrust required to maintain speed increase with speed decrease.

For a given mass, each aircraft has a minimum selectable speed and the maximum speed at a particular altitude. At the cruise altitude, there needs to be a safe margin in relation to these lowest and highest speeds, before the flight envelope protections activate.

At high altitude, reaching the aircraft's structural limit Mach number is almost impossible (except in a steep dive with the maximum thrust); therefore at the high altitude, flying at high Mach number should not be viewed as the biggest threat to the safety of flight. Conversely, flying too slow (below green dot) at the high altitude can lead to progressive reductions in speed until the protections are triggered. Should this speed reduction take place in a degraded law, it could lead to a loss of control due to stall. At and near the performance altitude limit of the aircraft, the range of available speed between green dot and MMO will be small. Speed decay at high altitude must be avoided as a result.

At lower altitudes (i.e. below the crossover altitude), too large a speed decay can similarly lead a non protected aircraft (i.e. flying in a degraded law) to enter a stall. Nevertheless, at a low altitude, the available envelope is greater and the thrust margin is much higher, thus providing flight crews a greater ability to safely control the airspeed and recover from a speed decay. On the other hand, at the low altitude, reaching VMO and VD is possible; therefore the high speed should be viewed indeed as a significant threat to the safety of flight.

When the AOA reaches the maximum value, the separation point moves further forward on the wing upper surface and almost total flow separation of the upper surface of the wing is achieved: this phenomenon leads to a significant loss of lift, referred to as a stall. Incidentally, stall is not a pitch issue and can happen at any pitch value.

任务 6　起飞
Task 6　Takeoff

 Contents

1) Takeoff

2) Takeoff speed

3) Factors affecting the takeoff speed

4) Takeoff distance

5) Factors affecting the takeoff distance

Learning Outcomes

1) Understand the process of aircraft takeoff

2) Master the measurement conditions of aircraft takeoff performance

3) Solve the dynamic problems in aircraft takeoff

4) Cultivate professional qualities of rigor, carefulness, and ability to express, coordinate, and communicate effectively

任务内容

1）起飞

2）起飞速度

3）起飞速度的影响因素

4）起飞距离

5）起飞距离的影响因素

任务目标

1）理解飞机起飞的过程

2）掌握飞机起飞性能的衡量条件

3）解决飞机起飞中的动力学问题

4）培养严谨、细心的职业素养，以及有效表达、协调和沟通的能力

For normal and practical aircraft, the process of taxiing from the runway, accelerating to the speed of lifting the front wheels, and raising the front wheels to an altitude of 50 ft above the takeoff surface, and reaching a safe takeoff speed is called takeoff. In the takeoff phase, the flight altitude of the aircraft is very low, and there is little room for maneuver in case of special circumstances. In addition, there is often low-level wind shear near the ground, so flight accidents are common in the takeoff phase. For pilots, mastering takeoff skills is one of the important subjects of flight training.

 课文

1. Takeoff

1. 起飞

Takeoff refers to the whole process that the aircraft starts taxiing from the takeoff line, accelerates to raising the nose wheels, continues to accelerate until the aircraft leaves the ground, and finally climbs over the safe altitude point (Fig. 6-14).

Takeoff can generally be divided into three stages: taxiing acceleration, pull up and acceleration climb.

起飞是指飞机从起飞线开始滑跑、加速到抬起前轮，继续加速直到飞机离开地面，最后爬升越过安全高度点的整个过程（图 6-14）。

起飞一般可分为三个阶段：地面滑跑加速、拉起和空中加速爬升。

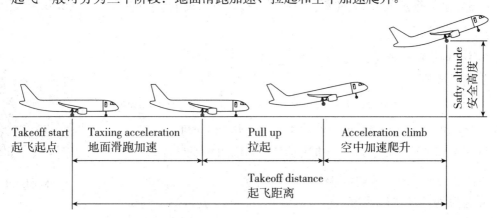

Fig. 6-14　Takeoff process
图 6-14　起飞过程

2. Takeoff Speed

2. 起飞速度

When an aircraft is in a takeoff taxiing, the speed when the lift is equal to the aircraft gravity is called the takeoff speed.

当飞机起飞前滑跑时，升力等于飞机重力时的速度称为起飞速度。

3. Factors Affecting the Takeoff Speed

3. 起飞速度的影响因素

Takeoff speed is related to the aircraft mass, air density and lift coefficient at takeoff.

(1) The greater the aircraft mass, the greater the takeoff speed.

(2) The smaller the air density, the greater the takeoff speed.

(3) The smaller the lift coefficient, the greater the takeoff speed.

(4) The higher the takeoff speed, the longer the distance from taxiing to takeoff, that is, the longer the distance of takeoff taxiing.

(5) The lift coefficient is related to the aircraft attitude and the high lift device.

(6) The takeoff attitude can increase the angle of attack and the use of high lift devices can reduce the takeoff speed and shorten the takeoff distance.

起飞速度与飞机起飞时的质量、空气密度和升力系数有关。

（1）飞机的质量越大，起飞速度越大。

（2）空气密度越小，起飞速度越大。

（3）升力系数越小，起飞速度越大。

（4）起飞速度越高，从滑跑到起飞的距离越长，即起飞滑跑距离越长。

（5）升力系数与起飞姿态和增升装置有关。

（6）起飞姿态可以增大迎角，使用增升装置可以减小起飞速度，缩短起飞滑跑距离。

Takeoff (1)

4. Takeoff Distance

4. 起飞距离

Takeoff distance is the horizontal distance from the beginning of takeoff taxiing to the safe altitude which the aircraft flies over.

起飞距离是从开始滑行到飞机飞越安全高度的水平距离。

5. Factors Affecting the Takeoff Distance

5. 起飞距离的影响因素

Takeoff (2)

The takeoff distance is related to the aircraft mass, the thrust of the engine, the atmospheric conditions, the use of the high lift device and the climb angle in the climbing stage.

起飞距离与飞机质量、发动机推力、大气条件、增升装置的使用和爬升阶段的爬升角度有关。

Takeoff (3)

 **New Words**

process	['prəʊses]	n.	过程，进程
taxi	['tæksi]	v.	（起飞前或降落后在地面上）滑跑
raise	[reɪz]	v.	提升，举起，提起
nose wheels	[nəʊz wiːlz]		前轮
stage	[steɪdʒ]	n.	阶段，段
high lift device	[haɪ lɪft dɪ'vaɪs]		增升装置

 Q&A

The following questions are for you to answer to assess the learning outcomes.

(1) Describe the definition of the takeoff.

(2) Describe the definition of the takeoff speed.

(3) Describe the definition of the takeoff distance.

(4) What factors affect the takeoff speed?

 Extended Reading

Takeoff Segments

The takeoff in a transport category aircraft is divided into four segments (Fig. 6–15). They are named, the first segment, second segment, third segment, and final segment. The takeoff segments are always presented with one engine inoperative.

The first segment starts from off the ground to the point where the landing gear is retracted. In this segment, the aircraft speed is V_{LOF}, the flaps are in takeoff configuration, and the engines must be at takeoff power. For a twin–engine aircraft, the climb gradient at this segment must be positive, and for a four–engine aircraft, it must be at least 0.5%.

The second segment starts from the gear retraction point to an altitude of 400 ft AGL (aboved ground level). The speed of the aircraft is V_2. The flaps are still in takeoff configuration, and the engines are at takeoff power. The climb gradient for a twin–engine aircraft is 2.4%, and that for a four–engine aircraft is 3.0%.

The third segment begins at 400 ft AGL. This is also known as the minimum acceleration altitude. At this segment, the aircraft's nose can be lowered and accelerated to flap retraction initiation speed to retract the flaps. Once flaps are retracted, the speed can be increased to minimum drag speed. The engines remain at takeoff power. There is no specific climb gradient requirement at this segment.

The final segment begins once the aircraft is clean (flaps up). At this segment, the minimum drag speed is reached, and the engine power is pulled to MCT (maximum continuous thrust). The final segment climb gradient for a twin-engine is 1.2%, and that for a four-engine is 1.7%. The takeoff segments end once the aircraft reaches 1,500 ft AGL, at which point the climb phase begins.

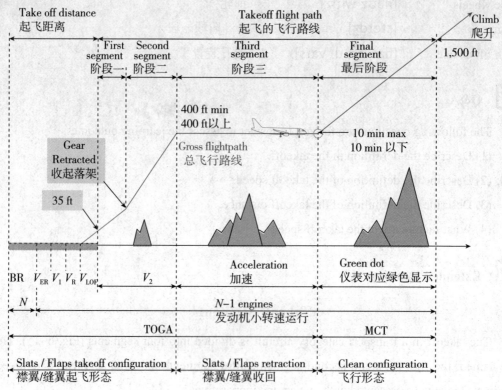

Fig. 6-15　Takeoff segments

图 6-15　起飞阶段

任务 7　着陆
Task 7　Landing

 Contents

1) Landing

2) Stages in the landing

3) Touching down speed

4) Factors affecting the touching down speed

5) Landing taxiing distance

6) Factors affecting the landing taxiing distance

Learning Outcomes

1) Understand the process of aircraft landing

2) Master the measurement conditions of aircraft landing performance

3) Solve the dynamic problems in aircraft landing

4) Cultivate professional qualities of rigor, carefulness, and ability to express, coordinate, and communicate effectively

任务内容

1）着陆
2）着陆的阶段
3）接地速度
4）接地速度的影响因素
5）着陆滑跑距离
6）着陆滑跑距离的影响因素

任务目标

1）理解飞机着陆的过程
2）掌握飞机着陆性能的衡量条件
3）解决飞机着陆中的动力学问题
4）培养严谨、细心，以及有效表达、协调和沟通的能力

Learning Guide

The landing process of an aircraft refers to the entire process of reducing its altitude and speed from an airborne state to the ground. When the aircraft is ready to land, the landing gear is lowered and begins to glide at a stable speed along a nearly oblique trajectory. When the aircraft glides to about 6-12 m above the ground, it pulls the joystick backwards, increases the angle of attack, and enters the flattening phase. Subsequently, the aircraft's trajectory gradually turns horizontaly, while its speed gradually decreased, and then entered the level flight phase. In the level flight phase, in order to maintain level flight while reducing speed, the aircraft's angle of attack continues to increase and

189

the flight speed further decreases. When the angle of attack increases to a point where it cannot be further increased, the aircraft gradually sinks under the influence of gravity and begins to enter the drifting phase. When the aircraft lands on the main wheel of the landing gear, it begins to roll on the ground. The pilot controls the brakes and deceleration devices to continue slowing down until the aircraft comes to a complete stop.

 课文

Landing
着陆

Landing is the process that the aircraft descends from a safe altitude, slows down in level flight, and taxis to a stop.

着陆是飞机从安全高度下降、平飞减速、滑跑至停稳的过程。

1. Stages Included in the Landing

1. 着陆包括的阶段

Landing generally includes five stages: glide, flare, level flight deceleration, touching down and landing taxiing (Fig. 6–16).

着陆一般包括下滑、拉平、平飞减速、接地和着陆滑跑五个阶段（图6–16）。

2. Touching Down Speed

2. 接地速度

Touching down speed is the speed at which the aircraft touches down during landing.

The lower the touching down speed, the safer the aircraft will land and the shorter the landing taxiing distance will be. Therefore, the lower the touching down speed, the better.

接地速度是指飞机在着陆过程中接地时的速度。

接地速度越小，飞机着陆越安全，着陆滑跑距离越短。因此，接地速度越低越好。

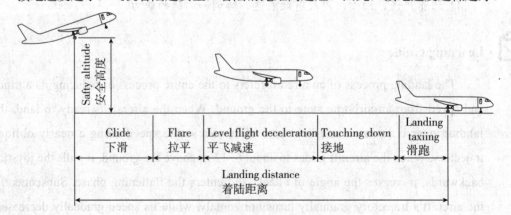

Fig. 6–16　Landing stages
图6–16　着陆过程

3. Factors Affecting the Touching Down Speed

3. 接地速度的影响因素

(1) Like the takeoff speed, the touching down speed is related to the aircraft mass, air density and the lift coefficient at landing.

(2) If the landing mass is too large, or the airport temperature is high, or the aircraft lands at a high altitude airport, the touching down speed will be excessive, the aircraft will be subjected to a large ground impact force when landing, damaging the landing gear and the body structure, and also making the landing distance too long, which results in some accidents, such as rushing out of the runway.

(3) For the safety of aircraft landing, the mass during landing shall not exceed the specified landing mass.

(4) The angle of attack at landing should take the maximum value, and the trailing edge flaps should be extended at the maximum angle during landing, so as to maximize the lift coefficient and reduce the touching down speed.

（1）与起飞速度一样，接地速度与飞机的质量、空气密度和着陆时的升力系数有关。

（2）如果着陆质量过大，或机场温度过高，或飞机在高海拔机场着陆，都会造成飞机接地速度过快。飞机接地时受到较大的地面撞击力，会损坏起落架和机身结构，也会使着陆距离过长，导致如冲出跑道等事故。

（3）为了飞机着陆的安全，着陆时的质量不得超过规定的着陆质量。

（4）接地时的迎角应为其最大值，且后缘襟翼应以最大角度伸出，以最大限度地增加升力系数，降低接地速度。

4. Landing Taxiing Distance

4. 着陆滑跑距离

The landing taxiing distance is the distance from the touching down point to the stop.

着陆滑跑距离是指从着陆点到停止滑跑的距离。

5. Factors Affecting the Landing Taxiing Distance

5. 着陆滑跑距离的影响因素

Landing (1)

The landing taxiing distance is related to the touching down speed and the deceleration rate.

The smaller the touching down speed, the faster the deceleration, and the shorter the landing taxiing distance.

In order to quickly reduce the aircraft speed during the landing taxiing, the spoiler for reducing lift and increasing drag should be opened after

Landing (2)

touching down, the brake and engine thrust reversers should be used.

着陆滑跑距离与着陆速度和刹车速度有关。

着陆速度越小，刹车速度越快，着陆滑行距离越短。

在着陆滑跑期间为了快速降低飞机速度，着陆后应打开减升增阻的扰流板，并使用刹车装置和发动机反推装置。

Landing (3)

 New Words

descends	[dɪ'sendz]	v.	下降，下斜，下倾
glide	[glaɪd]	v.	下滑，滑行，滑动，掠过，滑翔
flare	[fleə(r)]	v.	拉平
touching down	['tʌtʃɪŋ daun]		接地
airport	['eəpɔːt]	n.	机场，航空站，航空港
excessive	[ɪk'sesɪv]	adj.	过多的，过分的，过度的
landing gear	['lændɪŋ gɪə(r)]		起落架
accident	['æksɪdənt]	n.	事故，意外遭遇
rush	[rʌʃ]	v.	冲，迅速移动
exceed	[ɪk'siːd]	v.	超过（数量），超越
specified	['spesɪfaɪd]	v.	具体说明，明确规定
spoiler	['spɔɪlə(r)]	n.	扰流片，阻流板
brake	[breɪk]	n.	刹车，制动器
thrust reverser	[θrʌst rɪ'vɜːsə]		反推装置

 Q&A

The following questions are for you to answer to assess the learning outcomes.

(1) Describe the definition of landing.

(2) Describe the definition of touching down speed.

(3) Describe the definition of landing taxiing distance.

 Extended Reading

Landing Distance

The terms landing distance required (LDR) and landing distance available (LDA) routinely defined in aircraft landing performance documentation are not defined for fixed wing aeroplanes in ICAO SARPs. The ICAO definition for "landing distance" is usually taken as the basis for the determination of landing distance required which is calculated by taking into account the effect of various influencing factors, including prevailing surface conditions and the extent to which aircraft devices which are available to assist deceleration are deployed (Fig. 6-17).

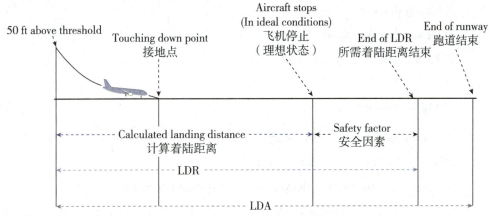

Fig. 6-17　Landing distance

图 6-17　着陆距离

Aircraft performance (LDR and landing speed) is calculated by the pilots using printed tables or a computer. This calculation takes account of the above factors, including the safety factor. It is assumed for these calculations that the aircraft will be at a specified altitude (normally 50 ft) crossing the runway threshold at the correct speed, and that aircraft handling will be in accordance with procedures detailed in the AFM and company SOPs.

Safety factors vary according to the aircraft type (turbo-jet or turbo-prop), the runway conditions (dry, wet or contaminated) and in pre-departure planning, whether the airfield is the destination or an alternate.

任务 8　匀速爬升和下滑
Task 8　Steady Climb and Descend

 Contents

1) Steady climb

2) Angle of climb

3) Climb rate

4) Rules of steady climb

5) Theoretical ceiling

6) Operating ceiling

7) Steady descend

8) Descend angle and lift drag ratio

9) Factors affecting the descend angle

Learning Outcomes

1) Master the forces of steady climb

2) Master the forces of steady descend

3) Solve the dynamics problems in steady climb and descend

4) Cultivate professional qualities of rigor, carefulness, and ability to express, coordinate, and communicate effectively

任务内容

1）等速爬升

2）爬升角

3）爬升率

4）等速爬升的规律

5）理论升限

6）实际升限

7）等速下滑

8）下滑角与升阻比

9）下滑角的影响因素

任务目标

1）掌握等速爬升的受力情况

2）掌握等速下滑的受力情况

3）解决等速爬升和下滑中的动力学问题

4）培养严谨、细心，以及有效表达、协调和沟通的能力

 ## Learning Guide

The climbing and descending procedures are part of the flight.

课文

1. Steady Climb

1. 等速爬升

Steady climb is that aircraft climbs at a steady speed along a straight line inclined upward to obtain flight altitude. Fig. 6-18 shows the forces of steady climb.

Steady climb is a balanced flight state.

等速爬升是指飞机沿向上倾斜的直线以稳定速度爬升以获得飞行高度。

等速爬升是一种平衡的飞行状态。图 6-18 所示为等速爬升时飞机的受力情况。

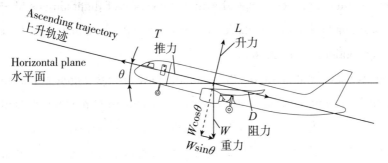

Fig. 6-18 Forces of steady climb
图 6-18 等速爬升受力情况

2. Angle of Climb

2. 爬升角

The angle of climb is the angle between the aircraft ascending trajectory and the horizontal plane, which is represented by θ.

$$\begin{cases} L=W \cdot \cos\theta \\ T=D+W \cdot \sin\theta \end{cases}$$

where L is lift, W is aircraft gravity, T is thrust, D is drag.

爬升角是飞机上升轨迹与水平面之间的夹角，用 θ 表示。

$$\begin{cases} L=W \cdot \cos\theta \\ T=D+W \cdot \sin\theta \end{cases}$$

式中，L 为升力；W 为飞机重力；T 为推力；D 为阻力。

3. Climb Rate

3. 爬升率

The climb rate is the altitude at which the aircraft climbs at a steady speed in a certain time.

爬升率是指飞机在一定时间内等速爬升的高度。

4. Rules of Steady Climb

4. 等速爬升的规律

(1) When climbing at a steady speed, the lift is less than the aircraft gravity, while the thrust

is greater than the drag.

(2) When the available thrust of the engine is greater than the required thrust (there is residual thrust), the aircraft can climb at a steady speed.

(3) The lighter the aircraft, the greater the residual thrust, the larger the climb angle of the aircraft.

(4) When climbing at a steady speed, the faster the speed of the aircraft, the larger the climbing angle, the larger the climbing rate, and the shorter the time it takes for the aircraft to climb to the same altitude, thus the better the ascending performance of the aircraft.

(5) When climbing at the steady speed, with the increase of flight altitude, the density of air gradually decreases, and the angle of attack must be increased to obtain a larger lift coefficient. In this way, the drag continues to increase.

With the increase of flight altitude, the available thrust of the engine continues to decrease, so that the residual thrust of the aircraft decreases rapidly and the climb rate gradually decreases.

（1）等速爬升时，升力小于飞机重力，而推力大于飞行阻力。

（2）当发动机的可用推力大于所需推力（存在剩余推力）时，飞机才能以稳定的速度爬升。

（3）飞机越轻，剩余推力越大，飞机的爬升角就可以越大。

（4）等速爬升时，飞机的速度越快，爬升角越大，爬升速率越大，飞机爬升到相同高度所需的时间越短，因此飞机的爬升性能越好。

（5）等速爬升时，随着飞行高度的增加，空气密度逐渐减小，必须增大迎角才能获得更大的升力系数。这样，飞行阻力会继续增加。

随着飞行高度的增加，发动机的可用推力不断减小，飞机的剩余推力迅速减小，爬升率逐渐减小。

5. Theoretical Ceiling

5. 理论升限

When the climb rate is equal to zero, the altitude of the aircraft is called the theoretical ceiling.

当爬升率等于零时，飞机的飞行高度称为理论升限。

6. Operating Ceiling

6. 实际升限

In practice, when the climb rate is less than a specified value, the altitude reached by the aircraft is called the ceiling (operating ceiling) (Fig. 6-19).

在实践中，当爬升率小于某一规定值时，飞机达到的高度称为升限（实际升限），如图 6-19 所示。

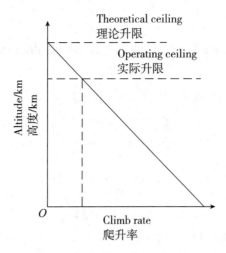

Fig. 6-19　Ceiling of the climbing

图 6-19　升限

7. Steady Descend

7. 等速下滑

Steady descend is the glide of an aircraft descending at a steady speed along a straight line without thrust.

The external forces acting on the aircraft is also balanced in steady descend (Fig. 6-20).

等速下滑是指飞机在零推力状态下沿直线以稳定速度下降。

等速下滑时作用在飞机上的外力平衡（图 6-20）。

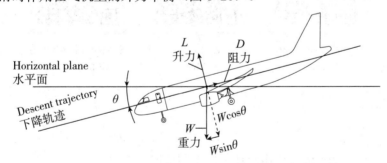

Fig. 6-20　Forces of steady descend

图 6-20　等速下滑受力情况

8. Descend Angle and Lift Drag Ratio

8. 下滑角与升阻比

The descend angle is the angle between the descent trajectory and the horizontal plane, which is represented by θ.

$$\begin{cases} L = W \cdot \cos\theta \\ D = W \cdot \sin\theta \end{cases}$$

Reciprocal of lift drag ratio is

$$\tan \theta = \frac{D}{L}$$

When the descent altitude is fixed, the larger the lift drag ratio, the longer the descent distance.

下滑角是下降轨迹与水平面之间的夹角，用 θ 表示。

$$\begin{cases} L = W \cdot \cos\theta \\ D = W \cdot \sin\theta \end{cases}$$

升阻比的倒数：

$$\tan \theta = \frac{D}{L}$$

当下降高度一定时，升阻比越大，下降距离越长。

9. Factors Affecting the Descend Angle

9. 下滑角的影响因素

The larger the lift drag ratio, the smaller the angle of descent.

The descend angle and distance is not affected by the aircraft gravity when gliding without thrust.

升阻比越大，下滑角越小。

零推力滑行时，下滑角和下滑距离不受飞机重力的影响。

Steady Climb and Descend (1) Steady Climb and Descend (2) Steady Climb and Descend (3)

 New Words

climb	[klaɪm]	v.	爬升，攀登
residual thrust	[rɪ'zɪdjuəl θrʌst]		剩余推力
rapidly	['ræpɪdlɪ]	adv.	迅速地，迅速
theoretical ceiling	[ˌθɪə'retɪkl 'siːlɪŋ]		理论升限
practical	['præktɪkl]	adj.	实际的，真实的
reciprocal	[rɪ'sɪprəkl]	n.	倒数
lift drag ratio	['lɪft dræg 'reɪʃʊ]		升阻比
gliding	['glaɪdɪŋ]	v.	滑行，滑动

Q&A

The following questions are for you to answer to assess the learning outcomes.

(1) Describe the definition of steady climb.

(2) What are the characteristics of the external forces when the aircraft climbs at a steady speed?

(3) Describe the definition of climb rate.

(4) Describe the definition of steady descent.

任务 9　水平转弯
Task 9　Level Turn

 Contents

1) Level turn and horizontal circling

2) Trajectory, speed, acceleration changes in level turn

3) Centripetal acceleration of level turn

4) Forces of level turn

5) Performance of level turn

6) Limits of increasing the bank angle

 Learning Outcomes

1) Master the method of level turn

2) Master the forces of level turn

3) Master the level turning performance of the aircraft

4) Solve the dynamic problems of aircraft level turn

5) Cultivate professional qualities of rigor, carefulness, and ability to express, coordinate, and communicate effectively

 任务内容

1）水平转弯和盘旋

2）水平转弯时的轨迹、速度、加速度变化

3）水平转弯时的向心加速度

4）水平转弯时的受力

5）水平转弯的性能

6）坡度角增大的限制

 任务目标

1）掌握飞机水平转弯的方法

2）掌握飞机水平转弯的受力情况

3）掌握飞机水平转弯的性能

4）解决飞机水平转弯中的动力学问题

5）培养严谨、细心的职业素养，以及有效表达、协调和沟通的能力

Learning Guide

The centripetal force of an aircraft turning in the air can only come from lift. Since the lift of an aircraft flying horizontally and straightly almost points vertically above the wing, and the lift is equal to gravity, in order to convert this lift into the centripetal force that makes the aircraft turn, it is necessary to tilt the aircraft to a certain angle by adjusting the aileron, so that part of the lift can be converted into the centripetal force that makes the aircraft turn.

 课文

1. Level Turn

1. 水平转弯

Level turn is a curvilinear motion in which an aircraft continuously changes its flight direction in a horizontal plane. Level turn refers to a maneuver with the course change angle less than $360°$.

A normal level turn is a circular maneuver without side slip and with steady speed. During normal level turn, the flight altitude of the aircraft does not change.

水平转弯是飞机在水平面内连续改变飞行方向的曲线运动。水平转弯是飞机航向变化角小于 $360°$ 的运动。

正常水平转弯是一种无侧滑且速度稳定的圆周运动。在正常水平转弯期间，飞机的飞行高度不变。

2. Horizontal Circling

2. 水平盘旋

Horizontal circling refers to a horizontal turn with a course change angle greater than 360°.

水平盘旋是飞机航向变化角大于 360° 的水平转弯。

3. Trajectory, Speed, Acceleration Changes in Level Turn

3. 水平转弯时的轨迹、速度、加速度变化

When the aircraft makes a level turn, the flight track changes from a straight line to a circle.

Although the flight speed does not change, the direction of the speed is constantly changing.

The change of speed direction indicates that the level turn of the aircraft has centripetal acceleration.

The acceleration direction is perpendicular to the tangent of the flight track and points to the center of the flight track.

当飞机水平转弯时，轨迹从直线变为圆。

虽然飞行速度没有改变，但速度的方向在不断变化。

速度方向的变化表明飞机的水平转弯具有向心加速度。

加速度方向垂直于轨迹的切线，并指向轨迹的中心。

4. Centripetal Acceleration of Level Turn

4. 水平转弯时的向心加速度

Centripetal acceleration is:

$$a_n = \frac{v^2}{R}$$

where v is the flight speed of the aircraft, R is the radius of the flight track.

The centripetal force is the product of the aircraft mass and the centripetal acceleration.

$$F_n = m \times a_n = \left(\frac{W}{g} \right) \times \left(\frac{v^2}{R} \right)$$

where m is the aircraft mass, W is the aircraft gravity, g is the acceleration of gravity.

向心加速度：

$$a_n = \frac{v^2}{R}$$

式中，v 是飞机的飞行速度；R 是转弯轨迹的半径。

向心力等于飞机质量和向心加速度的乘积。

$$F_n = m \times a_n = \left(\frac{W}{g} \right) \times \left(\frac{v^2}{R} \right)$$

式中，m 是飞机的质量；W 是飞机的重力；g 是重力加速度。

5. Forces of Level Turn

5. 水平转弯时的受力

(1) The forces on the aircraft during normal level turn are as Fig. 6-21.

$$\begin{cases} T=D \\ L \cdot \cos\gamma=W \\ L \cdot \sin\gamma=\dfrac{mv^2}{R}=\left(\dfrac{W}{g}\right)\times\left(\dfrac{v^2}{R}\right) \end{cases}$$

where T is the thrust of the engine, D is the drag of aerodynamic force, L is the lift of the aircraft, γ is the angle of roll (bank angle).

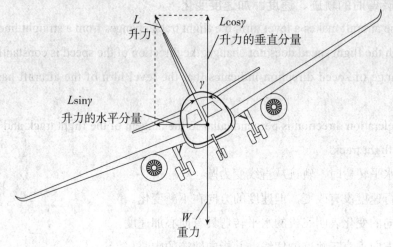

L
升力

Lcosγ
升力的垂直分量

Lsinγ
升力的水平分量

γ

W
重力

Fig. 6-21　Forces of level turn
图 6-21　水平转弯时的受力

(2) When the aircraft level turns, the thrust of the engine counteracts the drag and keeps the flight speed unchanged.

(3) The vertical component of lift counteracts the aircraft gravity, so that the altitude of the aircraft remains unchanged.

(4) The horizontal component of lift provides the centripetal force that makes the aircraft move in a curve, changes the direction of flight speed, and completes the level turns.

(5) When the aircraft is in level turn, the load factor is:

$$\begin{cases} L \cdot \cos\gamma = W \\ n_y = \dfrac{L}{W} = \dfrac{1}{\cos\gamma} \end{cases}$$

(6) When the aircraft is in level turn or horizontal circling, the load factor only depends on the bank angle γ.

(7) Since $\cos\gamma \leqslant 1$, the load factor $n_y \geqslant 1$, that is, the lift is always greater than the aircraft gravity.

(8) When turning, the greater the bank angle of the aircraft, the greater the lift required. For example, when the bank angle is 30°, $n_y = 1.15$, the lift is 1.15 times the aircraft gravity. For example, when the bank angle is 60°, $n_y = 2.0$, the lift is 2.0 times the aircraft gravity.

（1）在正常水平转弯时，飞机上的力如图 6-21 所示。

$$\begin{cases} T=D \\ L \cdot \cos \gamma = W \\ L \cdot \sin \gamma = \dfrac{mv^2}{R} = \left(\dfrac{W}{g} \right) \times \left(\dfrac{v^2}{R} \right) \end{cases}$$

式中，T 是发动机的推力；D 是空气动力的阻力；L 是飞机的升力；γ 是滚转角（坡度角）。

（2）当飞机水平转弯时，发动机的推力克服飞行阻力，并保持飞行速度不变。

（3）垂直方向的升力分量抵消了飞机的重力，因此飞机的高度保持不变。

（4）水平方向上的升力分量提供向心力，使飞机沿曲线运动，改变飞行速度方向，并完成水平转弯。

（5）当飞机水平转弯时，载荷系数为：

$$\begin{cases} L \cdot \cos \gamma = W \\ n_y = \dfrac{L}{W} = \dfrac{1}{\cos \gamma} \end{cases}$$

（6）当飞机水平转弯或水平盘旋时，载荷系数仅取决于坡度角 γ。

（7）因为 $\cos \gamma \leqslant 1$，所以载荷系数 $n_y \geqslant 1$，即升力总是大于飞机的重力。

（8）转弯时，飞机的坡度角越大，所需升力越大。再如，当坡度角为 30° 时，$n_y =$ 1.15，升力为飞机重力的 1.15 倍。例如，当坡度角为时 60°，$n_y = 2.0$，升力为飞机重力的 2.0 倍。

6. Performance of Level Turn

6. 水平转弯的性能

Turning radius R, and the time t required for one circle are two measurement standards of the level turning maneuverability of an aircraft.

The smaller the turning radius, the shorter the time it takes to circle, indicating the better level turning maneuverability of an aircraft.

Turning radius and circling time depend on flight speed and bank angle.

With the same bank angle, the higher the flight speed, the larger the turning radius, and the longer the time it takes to circle.

After the flight speed is determined, if the turning radius is to be reduced, the bank angle must be increased.

转弯半径 R 和盘旋一周所需时间 t 是飞机水平转弯机动性能的两个测量标准。

转弯半径越小，盘旋一周的时间越短，表明飞机的水平机动性能越好。

转弯半径和盘旋时间取决于飞行速度和坡度角。

在相同的坡度角下，飞行速度越高，转弯半径越大，盘旋一周的时间越长。

飞行速度确定后，如果要减小转弯半径，必须增大坡度角。

7. Limits of Increasing the Bank Angle

7. 坡度角增大的限制

However, increasing the bank angle will be limited by the following aspects.

(1) Firstly, increasing the bank angle will increase the load of the aircraft in the vertical axis direction, so the aircraft structure will bear a larger aerodynamic load, and the pilot will also bear a larger load. Therefore, increasing the bank angle is limited by the strength of the aircraft structure and the physiological conditions of the pilot.

(2) In addition, at a certain flight speed, increasing the bank angle must increase the angle of attack to increase the lift and avoid falling off the altitude, and the increase of the angle of attack is limited by the safety conditions, such as the critical angle of attack.

(3) Finally, when the bank angle increases, the angle of attack is increased and the lift is increased, the drag will also increase (Fig. 6-22). It is necessary to increase the thrust to keep the turning speed unchanged, which is limited by the available thrust of the engine at that altitude.

坡度角的增大将受以下几个方面的限制。

（1）首先，增加坡度角将增加飞机在垂直轴方向上的负载，因此飞机结构将承受更大的气动载荷，飞行员也将承受更大的负载。因此，增加坡度角受到飞机结构强度和飞行员生理条件的限制。

（2）此外，在一定的飞行速度下，增加坡度角必须增加迎角，以增加升力，避免掉高度，迎角的增加受到安全条件的限制，如临界迎角。

（3）最后，当坡度角增大时，迎角增大，升力增大，飞行阻力也增大。有必要增加推力以保持转弯速度不变（图 6-22），这受到该高度下发动机可用推力的限制。

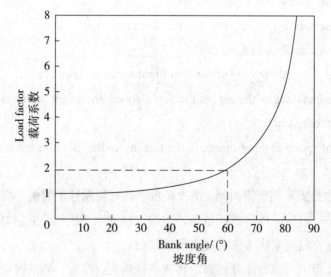

Fig. 6-22　Load factor and bank angle

图 6-22　载荷系数和坡度角

Level Turn (1)   Level Turn (2)   Level Turn (3)   Level Turn (4)   Level Turn (5)

 New Words

curvilinear	[ˌkɜːvɪˈlɪniə(r)]	*adj.*	曲线的
continuously	[kənˈtɪnjuəsli]	*adv.*	连续不断地
horizontal circling	[ˌhɒrɪˈzɒntlˈsɜːklɪŋ]		水平盘旋
course	[kɔːs]	*n.*	航向，课程
tangent	[ˈtændʒənt]	*n.*	切线，正切
track	[træk]	*n.*	轨迹，轨道，（人踩出的）小道
bank angle	[bæŋk ˈæŋgl]		坡度角，倾斜角
counteract	[ˌkaʊntərˈækt]	*v.*	抵制，抵消，抵抗
component	[kəmˈpəʊnənt]	*n.*	组成部分，成分
slop	[slɒp]	*n.*	坡度
measurement	[ˈmeʒəmənt]	*n.*	测量，度量，（某物的）尺寸
standards	[ˈstændədz]	*n.*	标准，水平，规格，规范
maneuverability	[məˌnuːvərəˈbɪlɪti]	*n.*	可操作性，机动性，可控性
radius	[ˈreɪdiəs]	*n.*	半径（长度）
overload	[ˌəʊvəˈləʊd]	*v.*	使超载，使负荷过重
physiological	[ˌfɪziəˈlɒdʒɪk(ə)l]	*adj.*	生理学的
falling	[ˈfɔːlɪŋ]	*v.*	落下，下落，掉落

 Q&A

The following questions are for you to answer to assess the learning outcomes.

(1) Describe the concept of level turn of the aircraft.

(2) Where does the acceleration direction point when the aircraft is in level turn?

(3) How is the external force on the aircraft balanced when the aircraft is in level turn?

(4) How to express the level turning maneuverability of the aircraft?

任务 10　侧滑
Task 10　Side Slip

 Contents

1) Side slip

2) Side slip angle

3) Left side slip and right side slip

4) Side slip in level turn

5) Inner side slip and outer side slip

6) Hazards of side slip

7) Methods to prevent side slip

8) Control of level turn of aircraft

Learning Outcomes

1) Master the characteristics and force of aircraft side slip

2) Master the generation conditions of aircraft side slip

3) Understand the hazards of aircraft side slip

4) Understand the preventive and corrective measures for aircraft side slip

5) Solve the dynamic problems in aircraft side slip

6) Cultivate professional qualities of rigor, carefulness, and ability to express, coordinate, and communicate effectively

 任务内容

1）侧滑

2）侧滑角

3）左侧滑与右侧滑

4）水平转弯时的侧滑

5）内侧滑与外侧滑

6）侧滑的危害

7）防止侧滑的方法

8）飞机水平转弯的操纵

任务目标

1）掌握飞机侧滑的特点和受力情况

2）掌握飞机侧滑的产生条件

3）了解飞机侧滑的危害

4）了解飞机侧滑的预防方法和纠正措施

5）解决飞机侧滑中的动力学问题

6）培养严谨、细心的职业素养，以及有效表达、协调和沟通的能力

 Learning Guide

Side slip is one of the common flight states when the aircraft encounters crosswind or undergoes yaw and roll during flight. In general, aircraft control should try to eliminate and reduce side slip as much as possible.

课文

Side Slip

侧滑

Side slip is the movement of an aircraft along its lateral axis.

During side slip, air flows from the side of the aircraft.

侧滑是飞机沿其横向轴线的运动。

侧滑期间，空气从飞机侧面流过。

1. Side Slip Angle

1. 侧滑角

The angle between the aircraft symmetry plane and the relative airflow is called the side slip angle (β) (Fig. 6–23).

飞机对称平面与相对来流之间的角度称为侧滑角（β），如图 6-23 所示。

2. Left Side Slip and Right Side Slip

2. 左侧滑与右侧滑

The air flows from the left nose of the aircraft is called left side slip, and the air flows from the right nose is called right side slip.

气流从飞机左机头流过称为左侧滑，从右机头流过称为右侧滑。

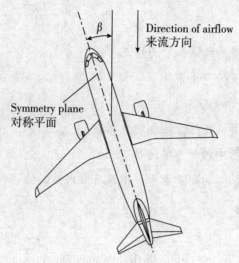

β

Direction of airflow
来流方向

Symmetry plane
对称平面

Fig. 6-23　Side slip

图 6-23　侧滑

3. Side Slip in Level Turn

3. 水平转弯时的侧滑

(1) When the aircraft is in level turn, the trajectory of the aircraft CG is an arc, and the moving speed of the CG follows the tangent direction of the arc.

(2) The aircraft body should rotate around the CG to the turning side to turn the nose to the tangent direction of the arc, that is, to align with the airflow.

(3) If the aircraft body rotates improperly, it will cause the aircraft symmetry plane to deviate from the flight track and thus produce side slip.

（1）当飞机水平转弯时，飞机重心的轨迹是一条圆弧，重心的移动速度沿该圆弧的切线方向。

（2）飞机机体应围绕重心向转弯侧转动，以将机头转向圆弧的切线方向，即对准来流。

（3）如果机体旋转不当，将导致飞机对称面偏离飞行轨迹，从而产生侧滑。

4. Inner Side Slip and Outer Side Slip

4. 内侧滑与外侧滑

The air flows from inside of the turning arc is called inner side slip, and the air flows from the outside is called outer side slip (Fig. 6-24).

气流从转弯弧内侧流过称为内侧滑，从外侧流过称作外侧滑（图 6-24）。

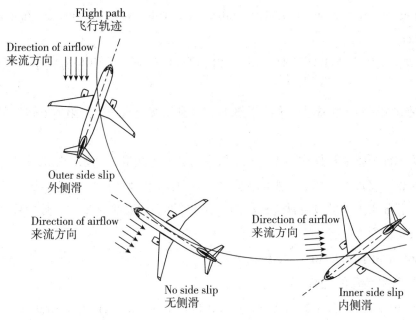

Fig. 6-24　Inner and outer side slip

图 6-24　内侧滑和外侧滑

5 Hazards of Side Slip

5. 侧滑的危害

The lateral force generated by side slip will cause the external forces and moment on the aircraft to change, making the aircraft deviate from the expected flight track.

In order to maintain the normal level turn and circling of the aircraft, it is necessary to prevent the aircraft from side slipping.

侧滑产生的侧向力将引起飞机上的外力和外力矩发生变化，使飞机偏离预期的飞行轨迹。

为了保持飞机的正常水平转弯和盘旋，必须防止飞机侧滑。

6. Methods to Prevent Side Slip

6. 防止侧滑的方法

In order to prevent the aircraft from side slip when turning horizontally, the rudder should also be deflected to take advantage of the lateral force generated by the rudder deflection.

Another method is to open the spoilers on the side of the upper aileron, and use the drag generated by the spoilers to deflect the aircraft nose and align it with the relative airflow.

If the rudder or spoilers are not operated properly, side slip will occur when the aircraft is in level turn. For example, if the rudder is pushed too much, the aircraft nose deflection will be too much, which will produce lateral side slip, resulting in inward lateral force, and lead to unexpected flight state of the aircraft. If the rudder is not pushed enough, it will produce outer lateral sliding, resulting in outward lateral force, which will lead to unexpected flight state of the aircraft either.

为了防止飞机在水平转弯时侧滑，方向舵也应偏转，以利用方向舵偏转产生的侧向力。

另一种方法是打开上副翼一侧的扰流板，并利用扰流板产生的阻力偏转副翼，使其与来流对齐。

如果方向舵或扰流板操作不当，水平转向时会发生侧滑。例如，如果推动方向舵的力过大，机头偏转将过大，从而产生侧向滑动，导致向内的侧向力，从而导致飞机处于意外的飞行状态。如果方向舵没有被充分推动，它将产生外部横向滑动，导致向外的横向力，这将导致飞机的意外飞行状态。

7. Control of Level Turn of Aircraft

7. 飞机水平转弯的操纵

(1) To operate the aircraft to make a level turn, it is necessary to operate the ailerons to toll the aircraft to the turning side, to obtain the centripetal force required for maneuvering, to generate a bank angle and maintain it during the turning process.

(2) Lift can produce a horizontal component and provide centripetal force for aircraft level turn.

(3) While keeping the flight speed constant, at the same time, pull the control column backward to increase the angle of attack in order to increase the lift, of which vertical component counteracts the aircraft gravity to prevent the aircraft from falling off the altitude when in level turn.

(4) The increase of the angle of attack will not only increase the lift, but also increase the drag.

(5) In order to maintain the flight speed, the engine thrust should also be increased to counteract the increased drag, so as to meet the requirement that the thrust is equal to the drag.

(6) At the same time, the rudder should be deflected or the spoiler should be extended upward to prevent the aircraft from side slip.

(7) Therefore, in order to keep the aircraft in level turn without side slip, it is necessary to coordinate the ailerons, elevators and rudders. In addition, it should cooperate with the throttle control of the engine to maintain the appropriate thrust.

（1）操纵飞机进行水平转弯，需要操纵副翼使飞机转向一侧，以获得机动所需的向心力，产生坡度角，并在转弯过程中保持坡度角不变。

（2）升力可以在水平方向上产生分量，并为飞机转弯提供向心力。

（3）在保持飞行速度不变的同时，向后拉动驾驶杆以增加迎角，增加升力，使升力的垂直分量抵消飞机的重力，防止飞机在水平转弯时掉高度。

（4）迎角的增加不仅会增大升力，还会增大阻力。

（5）为了保持飞行速度不变，还应增加发动机推力以抵消增大的阻力，以满足推力等于阻力的要求。

（6）同时，方向舵应当偏转或扰流板上偏，以防飞机侧滑。

（7）因此，为了使飞机水平转弯而不发生侧滑，必须协调副翼、升降舵和方向舵。此外，还应与发动机的油门操纵配合，以保持适当的推力。

Side Slip (1)   Side Slip (2)   Side Slip (3)   Side Slip (4)

 New Words

side slip	[said slɪp]	侧滑，侧滑移
arc	[ɒk]	*n.* 弧，弧形
tangent	['tændʒənt]	*adj.* 切线的，正切的
align	[ə'laɪn]	*v.* 排列，校准，排整齐
improperly	[ɪm'prɒpəli]	*adv.* 不正确地，不适当地
deviate	['diːvieɪt]	*v.* 背离，偏离
expected	[ɪk'spektɪd]	*adj.* 预料的，预期的
deflect	[dɪ'flekt]	*v.* 偏转，转移
push	[pʊʃ]	*v.* 推，推动（人或物），移动
centripetal	[ˌsentrɪ'piːtl]	*adj.* 向心的
control column	[kən'trəʊl 'kɒləm]	驾驶杆
coordinate	[kəʊ'ɔːdɪneɪt]	*n.* 坐标，（颜色协调的）配套服装
appropriate	[ə'prəʊpriət]	*adj.* 适当的，合适的

 Q&A

The following questions are for you to answer to assess the learning outcomes.

(1) Describe the definition of side slip.

(2) Describe the definition of side slip angle.

(3) How did the side slip of the aircraft occur?

(4) What methods can prevent side slip?

211

任务 11 增升装置
Task 11 High-Lift Device

Contents

1) Purpose of high-lift device

2) Importance of high-lift device

3) High-lift principle

4) Trailing edge flap

5) Leading edge flap

6) Leading edge slats and their functions

7) Working principle of slats

8) Control of leading edge slats and trailing edge flap

9) High-lift device for controlling boundary layer

Learning Outcomes

1) Master the high-lift principle of high-lift devices

2) Master the common types of high-lift devices and their working principles

3) Master the problems when using high-lift devices

4) Solve the dynamic problems of the aircraft in the use of high-lift devices

5) Cultivate professional qualities of rigor, carefulness, and ability to express, coordinate, and communicate effectively

任务内容

1）增升装置的目的

2）增升装置的重要性

3）增升原理

4）后缘襟翼

5）前缘襟翼

6）前缘缝翼及其功能

7）缝翼的工作原理

8）前缘缝翼和后缘襟翼的操纵

9）控制附面层的增升装置

 任务目标

1）掌握增升装置的增升原理

2）掌握常见的增升装置类型及其工作原理

3）掌握增升装置使用时的问题

4）解决飞机在增升装置使用中的动力学问题

5）培养严谨、细心的职业素养，以及有效表达、协调和沟通的能力

 Learning Guide

When the aircraft is cruising normally, due to its fast speed, it can provide sufficient lift to ensure the aircraft's flight altitude. When an aircraft takes off and lands, for safety reasons, its flight speed is usually relatively slow, resulting in less lift. Moreover, it is adopted as a certain sweep angle design in modern aircraft to reduce the lift performance during takeoff and landing. Therefore, during the takeoff and landing stages of an aircraft, a certain lift increasing device is usually used to enhance the lift of the aircraft during this stage. It is called as high-lift device.

课文

Purpose of High-lift Device:

增升装置的目的：

The purpose of high-lift devices is to obtain a greater lift when the aircraft flies slowly, reduce the takeoff speed of the aircraft, reduce the landing speed of the aircraft, improve aircraft takeoff and landing performance, improve the safety of aircraft during takeoff and landing, increase the critical angle of attack to prevent the aircraft from stalling at a large angle of attack, and thereby improve the lift coefficient.

增升装置的目的是在飞机飞行速度较慢时，获得更大的升力；降低飞机的起飞速度；降低飞机的着陆速度；提高飞机的起飞和着陆性能；提高飞机起飞和着陆的安全性；增大临界迎角，防止飞机在大迎角情况下失速，还可以提高升力系数。

Importance of High-lift Device:

增升装置的重要性：

With the development of high speed modern aircraft, the takeoff and landing speed of these large aircraft will become faster.

Since large mass of these aircraft, the lift required is larger, and the aircraft is also required to maintain a higher flight speed when taking off and landing. And the wings of high speed aircraft do not perform well at low speeds, such as thin airfoils, swept wings, etc. To use wings with poor low speed performance to achieve a certain lift at low speeds, a higher flight speed is inevitable. Therefore, high-lift device is more important for improving the safety of modern civil transport aircraft during take off and landing.

随着现代高速飞机的发展，这些大型飞机的起飞和着陆速度将变得更快。

由于这些飞机质量较大，所需升力较大，这也要求飞机在起飞和着陆时保持较高的飞行速度。而高速飞机的机翼在低速时性能并不好，如薄翼型、后掠翼等。要使用低速性能不好的机翼在低速时获得一定的升力，必然需要更高的飞行速度。因此，增升装置对于提高现代民用运输机起飞和着陆的安全性来说就更为重要。

Principle of Increasing Lift:

增加升力的原理：

According to the lift formula, the principles of increasing lift can be obtained.

$$L=C_L \cdot \frac{1}{2}\rho v^2 \cdot S$$

Principle 1: Change the cross section shape of the wing and increase the camber of the wing. Increasing the camber of the wing can speed up the flow speed of the upper wing surface and increase the negative pressure value of the upper wing surface, so as to improve the lift coefficient. However, increasing the camber of the wing will also increase the differential pressure drag and reduce the critical angle of attack.

Principle 2: Increase the wing area. Increasing the wing area can increase the lift, but it will also increase the drag.

Principle 3: Control the boundary layer on the wing to delay the separation of airflow. Controlling the boundary layer is to use some aerodynamic devices on the aerodynamic surface to continuously input dynamic energy into the boundary layer, or absorb and blow out the boundary layer.

These methods can accelerate the airflow in the boundary layer to reduce the thickness of the boundary layer, and delay the separation of the boundary layer.

根据升力公式，可以得出增加升力的原理。

$$L=C_L \cdot \frac{1}{2}\rho v^2 \cdot S$$

方法 1：改变机翼截面的形状，增加翼型的弯度。增加翼型弯度可以加快上翼面的气流流速，增加上翼面的负压值，从而提高升力系数。然而，增加翼型弯度也会增加压差阻力并减小临界迎角值。

方法 2：增加机翼面积。增加机翼面积可以增加升力，但同时也会增加阻力。

方法 3：控制机翼上的附面层以推迟气流分离。控制附面层就是在气动表面上使用一些气动装置，将动态能量连续输入附面层，或吸取、吹除附面层。

这些方法可以加速附面层内气流的流动以减小附面层的厚度，并推迟附面层的分离。

1. Trailing Edge Flap

1. 后缘襟翼

The trailing edge flap has many forms, such as plain flap, split flap, slotted flap, fowler flap and so on.

后缘襟翼有多种形式，如简单襟翼、分裂式襟翼、开缝式襟翼、后退开缝式襟翼等。

1) Plain Flap

1）简单襟翼

The plain flap is a small airfoil installed on the trailing edge of the wing that can rotate around the rotating shaft (Fig. 6-25). It closes as part of the trailing edge of the wing when not in use, and it deflects downward around its shaft when in use. Its high-lift principle is to change the shape of the wing airfoil to increase the curvature of the wing to accelerate the airflow on the upper wing and decelerate the airflow on the lower wing, hence to increase the pressure difference between the upper and lower wings, and thus increase the lift.

简单襟翼是一种安装在机翼后缘的小翼型，可绕转轴转动（图 6-25）。它在不使用时，闭合成为机翼后缘的一部分，使用时，围绕轴向下偏转。它的增升原理是改变机翼剖面的形状，以增大机翼弯度，加速上翼面的气流，减速下翼面的气流，从而增大上下翼面之间的压差，增加升力。

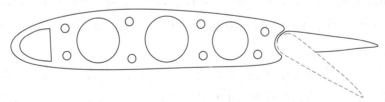

Fig. 6-25　Plain flap

图 6-25　简单襟翼

2) Split Flap

2）开裂式襟翼

The split flap is a plate that can rotate around the shaft on the lower surface of the trailing edge of the wing (Fig. 6-26). It retracts and clings to the lower surface of the wing, and becomes a part of the lower surface of the trailing edge of the wing when not in use, and it opens downward around the shaft when in use. The principle of it is to increase the curvature of the wing to increase lift. When it is opened, a low pressure area is formed between the flap and the upper surface of the trailing edge, attracting the airflow on the upper surface to accelerate, thus increasing the pressure difference between the upper and lower surfaces and increasing the lift.

　　分裂式襟翼是在机翼后缘下表面的一块可绕轴旋转的板件（图6-26）。在不使用时，它缩回并紧贴在机翼下表面，成为机翼后缘下表面的一部分；在使用时绕轴向下打开。其原理是增加翼型弯度以增加升力：当襟翼打开时，在襟翼和机翼后缘上表面之间形成一个低压区域，吸引上表面上的气流加速，从而增加上下翼面之间的压差，增加升力。

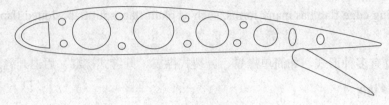

Fig. 6-26　Split flap
图 6-26　分裂式襟翼

3)Slotted Flap
3）开缝式襟翼

Slotted flap is improved on the basis of simple flap. It moves the rotating shaft from the center of the flap leading edge to the lower surface of the flap leading edge (Fig. 6-27). The flap opens downward around the rotation shaft, which not only increases the curvature of the wing, but also forms a gap between the leading edge of the flap and the rear of the wing, so that the high pressure airflow on the lower wing surface accelerates to the upper wing surface through the gap, transports dynamic energy to the boundary layer of the upper wing surface to prevent airflow separation, and greatly improves the lift increasing effect of the flap. Since it uses two high-lift principles of increasing wing curvature and controlling boundary layer, the high-lift effect is better.

　　开缝式襟翼是在简单襟翼的基础上改进的。它将转轴从襟翼前缘中心移动到襟翼前缘下表面（图6-27）。襟翼绕转轴向下打开，这种方式不仅增加了翼型弯度，而且在襟翼前缘和机翼后部之间形成了一个间隙，使下翼面上的高压气流通过该间隙加速流到上翼面，将动力能量传输到上翼面的附面层，以防气流分离，大大提高襟翼的增升效果。由于采用了增加翼型弯度和控制附面层两种增升原理，增升效果更好。

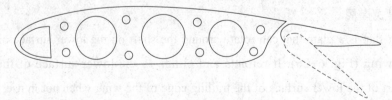

Fig. 6-27　Slotted flap
图 6-27　开缝式襟翼

4) Fowler Slotted Flap
4）后退开缝式襟翼

When the fowler slotted flap works, the flap deflects downward while retreating. While backward deflection, the fowler slotted flap forms a gap between the leading edge of the flap and the rear of the wing. The high-pressure airflow on the lower wing accelerates to the upper wing through the gap, accelerating the flow of the boundary layer on the upper wing and preventing the separation of airflow. Since this flap adopts three high-lift principles of increasing the camber of the wing, increasing wing area and controlling boundary layer, the high-lift effect is particularly good (Fig. 6-28). Some high-performance aircraft with relatively small thickness of wing airfoil tend to use this kind of high-lift device.

当后退开缝式襟翼工作时，襟翼在后退时向下偏转。当后退偏转时，后退开缝式襟翼在襟翼前缘和机翼后部之间形成间隙，下翼面的高压气流通过间隙加速流到上翼面，加速上翼面附面层的流动，防止气流分离。由于这种襟翼采用了增加翼型弯度、增加机翼面积和控制附面层的三种增升原理，增升效果特别好（图 6-28）。一些机翼翼型相对厚度较小的高性能飞机倾向于使用这种增升装置。

Fig. 6-28　Fowler slotted flap
图 6-28　后退式开缝襟翼

5) Fowler Double/Triple Slotted Flap
5）后退双 / 三开缝式襟翼

If only one gap is opened, the airflow to the upper wing surface is limited. When the flap deflects to a certain extent, the airflow will still separate, and the flap will vibrate. To solve this problem, one or two small airfoils are installed on the leading edge of the flap. The small airfoil and the main flap are separated to form two or three slots (gaps), through which more high-pressure airflow is accelerated from the lower wing surface to the upper wing surface. In this way, when the flap deflects at a larger angle, the airflow separation will not occur, and a better high-lift effect can be obtained. At present, this type of trailing edge flap is mostly used in large high-speed civil transport aircraft in order to obtain a higher high-lift effect (Fig. 6-29).

如果仅打开一个间隙，则流向上翼面的气流受限。当襟翼偏转到一定程度时，气流仍将分离，襟翼将发生振动。为了解决这个问题，在襟翼的前缘安装一个或两个小翼面：小翼面和主襟翼分开形成两个或三个缝（间隙），更多的高压气流通过这两个或三个开缝从下翼面加速到上翼面。这样，当襟翼偏转较大角度时，不会发生气流分离，可以获得更好的增升效果。目前，这种后缘襟翼主要用于大型高速民用运输机，以获得更好的增升效果（图 6-29）。

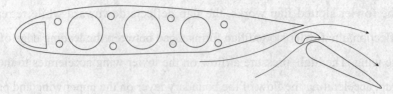

Fig. 6-29 Fowler double/triple slotted flap
图 6-29 后退式双 / 三开缝襟翼

Usage of Trailing Edge Flap is as below.

后缘襟翼的使用方法如下。

All kinds of flaps can significantly improve the lift coefficient of the wing, among which the fowler slotted flap has the best high–lift effect. However, the trailing edge flap not only improves the lift coefficient of the wing, but also increases the drag coefficient of the wing. When the flap extension angle is small, the percentage of drag increase is lower than that of lift, which is suitable for the takeoff procedure that requires more lift and less drag. When the flap extension angle is large, the percentage of drag increase is almost the same as the lift, which is applicable for the landing procedure that requires both lift and drag are much as possible. Therefore, although trailing edge flaps are both used during takeoff and landing, the methods of them are different. The extension of the trailing edge flap is at a moderate angle of about 20° when taking off. By contrast, the extension of the trailing edge flap is about 40° when landing.

When using trailing edge flaps to improve the lift coefficient, the critical angle of attack decreases. In this way, when taking off and landing at a large angle of attack, the use of trailing edge flaps can easily cause the aircraft to stall, so the trailing edge flaps of modern large transport aircraft are used together with leading edge flaps, leading edge slats and other lift increasing devices.

各种襟翼都能显著提高机翼的升力系数，其中后退开缝式襟翼的增升效果最好。虽然，后缘襟翼提高了机翼的升力系数，但是也增加了机翼的阻力系数。当襟翼下放角度较小时，阻力增加的百分比低于升力增加的百分比，这适用于需要更多升力和更少阻力的起飞状态。当襟翼下放角度较大时，阻力增加的百分比几乎与升力相同，这适用于要求升力和阻力都尽可能大的着陆状态。因此，虽然起飞和着陆时都使用后缘襟翼，但使用的方法不同。起飞时后缘襟翼的下放角度适中约为 20°，而着陆时后缘襟翼的下放角度约为 40°。

当使用后缘襟翼提高升力系数时，临界迎角减小。这样，当以大迎角起飞和着陆时，使用后缘襟翼容易导致飞机失速，因此现代大型运输机的后缘襟翼都是与前缘襟翼、前缘缝翼等其他增升装置一起使用。

2. Leading Edge Flap

2. 前缘襟翼

The leading edge flap is a high-lift device installed on the leading edge of the wing.

(1) Krueger flaps are located at the leading edge of the wing (Fig. 6-30). It is a panel on the lower surface of the leading edge of the wing. It clings to the lower surface of the wing to form the outer surface when not in use, and the actuator extends outward and pushes the Krueger flap to rotate around the front rotating shaft to open forward and downward when used.

(2) The retractable leading edge flap is a controllable leading edge of the wing (Fig. 6-31). It keeps the shape of the leading edge when not in use, and driven by the actuator, the entire leading edge slides downward to form a low wing leading edge when in use.

These two flaps are generally used in high-speed aircraft.

前缘襟翼是安装在机翼前缘的增升装置。

（1）克鲁格襟翼位于机翼前缘（图 6-30）。它是机翼前缘下表面的一块面板。不使用时，紧贴机翼前缘下表面形成机翼外翼面；使用时，作动筒向外伸出，推动克鲁格襟翼绕前转轴旋转，向前下方打开。

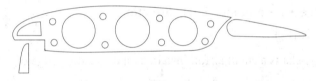

Fig. 6-30　Krueger flap
图 6-30　克鲁格襟翼

（2）下垂式前缘襟翼是可控的机翼前缘（图 6-31）。它在不使用时保持前缘的形状，使用时，在作动筒的驱动下，整个前缘向下滑动，形成低垂的机翼前缘。

这两类襟翼通常用于高速飞机。

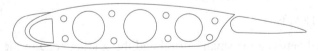

Fig. 6-31　Retractable leading edge flap
图 6-31　下垂式前缘襟翼

Usage of Leading Edge Flap is as below.

前缘襟翼的使用方法如下。

The wings of high-speed aircraft adopt airfoils with sharp leading edges and relatively small thickness.

When a high-speed aircraft flies at a certain angle of attack, since there is no smooth airflow on the upper surface of the leading edge, the flow separation will occur when the airflow is frustrated at the leading edge, which will greatly reduce the lift coefficient of the wing. At this time, if the leading edge flap is extended or the Krueger flap is opened, the angle between the leading edge and the relative airflow can be reduced, so that the airflow flows smoothly through the upper wing surface without separation.

Especially when the aircraft uses the trailing edge flap which will cause airflow separation and greatly reduce the high−lift effect. If the leading edge flap is extended at the same time, the air separation at the leading edge of the wing can be eliminated and the high−lift effect of the trailing edge flap can be improved.

高速飞机的机翼采用前缘尖锐、相对厚度较小的翼型。

当高速飞机以一定迎角飞行时，由于前缘上表面没有形成平滑的流道，当气流在前缘受阻时，将出现气流分离，这将大大降低机翼的升力系数。此时，如果前缘襟翼下垂或克鲁格襟翼打开，前缘与相对来流之间的夹角可以减小，从而使气流顺利流过上翼面，而不会分离。

当飞机使用后缘襟翼时，会导致气流分离并大大降低增升效果。如果同时使用前缘襟翼，则可以消除机翼前缘的气流分离，并提高后缘襟翼的增升效果。

3. Leading Edge Slats and and Their Functions
3. 前缘缝翼及其功能

The leading edge slat is a small airfoil installed on the leading edge of the wing.

The leading edge slats have two functions: The first is to increase the critical angle of attack and reduce the stall speed of the aircraft. The second is to increase the maximum lift coefficient.

前缘缝翼是安装在机翼前缘的小翼面。

前缘缝翼有两个功能：第一个功能是增加飞机的临界迎角，降低失速速度。第二个功能是增加最大升力系数。

Working Principle of Slats is as below.

缝翼的工作原理如下。

A gap is formed between the small airfoil and the leading edge of the wing when the slat works (Fig. 6−32). The high pressure airflow on the lower surface of the wing is accelerated to the upper surface through the gap, brings dynamic energy to the boundary layer of the upper wing to eliminate the vortex and increase the lift, hence to delay the separation of the airflow, and avoids stall at high angles of attack.

缝翼工作时，小翼面和机翼前缘之间形成间隙（图 6-32）。下翼面的高压气流通过间隙加速流到上翼面，将动力能量带到上翼面附面层，以消除涡流并增加升力，从而延迟气流分离，并避免在大迎角下失速。

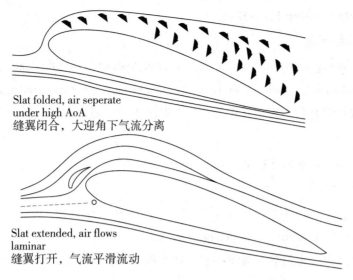

Slat folded, air seperate
under high AoA
缝翼闭合，大迎角下气流分离

Slat extended, air flows
laminar
缝翼打开，气流平滑流动

Fig. 6-32 Airflow on the wing with or without working slats

图 6-32 缝翼打开前 / 后机翼上的气流情况

1) Fixed Leading Edge Slats

1）固定式前缘缝翼

The fixed leading edge slats use ribs to fix the small airfoil on the leading edge of the wing (Fig. 6-33). With or without leading edge slats, a fixed gap is formed between the small airfoil and the leading edge of the wing. This kind of leading edge slat will increase the drag when the speed increases. It is not widely used on modern aircraft.

固定式前缘缝翼利用肋板，将小翼面固定在机翼前缘上（图 6-33）。无论是否使用前缘缝翼，小翼面和机翼前缘之间都会形成固定间隙。当速度增加时，这种前缘缝翼会增加空气阻力。

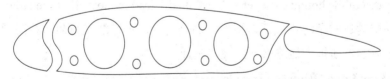

Fig. 6-33 Fixed leading edge slats

图 6-33 固定式前缘缝翼

2) Movable Leading Edge Slats

2）可动式前缘缝翼

Movable leading edge slats can be automatic or controllable configuration, on which the small airfoil is connected with the leading edge through a sliding mechanism.

可动式前缘缝翼可以是自动式构型或可控式配置，小翼面通过滑动机构与前缘连接。

a. Automatic Leading Edge Slats

a. 自动式前缘缝翼

Automatic leading edge slats use the aerodynamic load acting on the leading edge to extend or retract the small airfoil, which is mostly used in low altitude and low speed aircraft.

自动式前缘缝翼是利用作用在前缘的气动载荷来伸出或收回小翼面，主要用于低空低速飞机。

b. Controllable Leading Edge Slats

b. 可控式前缘缝翼

Controllable leading edge slats usually work automatically with the trailing edge flaps to prevent the aircraft from stalling at high angles of attack.

可控式前缘缝翼通常与后缘襟翼一起自动工作，以防飞机在大迎角下失速。

c. Control of Leading Edge Slats and Trailing Edge Flap

c. 前缘缝翼和后缘襟翼的操纵

A handle is used to control the leading edge slats and the trailing edge flap in some large transport aircraft. When extended, the leading edge slats are deployed prior to the trailing edge flaps. When retracted, the leading edge slats are retracted prior to the trailing edge flaps.

在一些大型运输机中，操纵前缘缝翼和后缘襟翼使用同一个手柄。伸出时，前缘缝翼在后缘襟翼伸展之前伸出。收回时，前缘缝翼在后缘襟翼收回之前收回。

4. High-Lift Device for Controlling Boundary Layer

4. 控制附面层的增升装置

The purpose of controlling the boundary layer is to prevent or delay the separation of the boundary layer, which will increase the drag and reduce the lift, also the continuous expansion of the separation area of the boundary layer will eventually lead to the stall of the aircraft.

控制附面层的目的是防止或延迟附面层的分离，这将增加飞行阻力并降低升力，而且附面层分离区域的不断扩大最终将导致飞机失速。

1) Boundary Layer Blowing Device

1）附面层吹除装置

The boundary layer blowing device (Fig. 6-34) blows the high pressure air out of the upper surface of the wing and brings dynamic energy into the boundary layer to accelerate the airflow speed and delay the separation of the boundary layer.

附面层吹除装置（图6-34）将高压气流吹出上翼面，并将动能输入附面层，以加速气流速度并延迟附面层的分离。

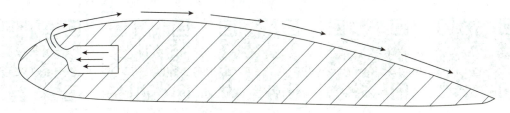

Fig. 6-34   The boundary layer blowing device
图 6-34   附面层吹除装置

2) Boundary Layer Suction Device
2）附面层吸取装置

The boundary layer suction device (Fig. 6-35) sucks the boundary layer through the gap on the upper surface of the wing to reduce the thickness of the boundary layer, and speed up the airflow in the boundary layer.

附面层吸取装置（图6-35）通过上翼面的间隙吸取附面层，以减小附面层的厚度，并使附面层内的气流加速。

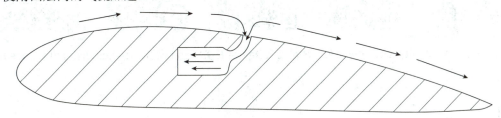

Fig. 6-35   Boundary layer suction device
图 6-35   附面层吸收装置

3) Vortex Generator
3）涡流发生器

The vortex generator continuously brings energy from the external airflow into the boundary layer to accelerate the flow speed of the boundary layer, and delay the separation of the airflow. The critical angle of attack can be enhanced and the lift coefficient can be increased when the vortex generator is used in low speed aircraft. Shock wave separation can be delayed when it is used in high speed aircraft.

涡流发生器不断地将外部气流的能量输入附面层，以加速附面层内气流的流动速度，并延迟气流的分离。当低速飞机使用涡流发生器时，可以提高临界迎角，增加升力系数。当用于高速飞机时，它可以推迟激波分离。

High Lift Device (1)

High Lift Device (2)　High Lift Device (3)　High Lift Device (4)　High Lift Device (5)　High Lift Device (6)

High Lift Device (7)　High Lift Device (8)　High Lift Device (9)　High Lift Device (10)　High Lift Device (11)

High Lift Device (12)　High Lift Device (13)　High Lift Device (14)　High Lift Device (15)　High Lift Device (16)

 New Words

high-lift device		增升装置
improve	[ɪm'pruːv]	v. 改进，改善
perform	[pə'fɔːm]	v. 表演，执行，履行
thin	[θɪn]	adj. 薄的，细的，瘦的，稀少的
achieve	[ə'tʃiːv]	v. 实现，完成，（凭长期努力）达到
inevitable	[ɪn'evɪtəbl]	adj. 不可避免的，不能防止的
formula	['fɔːmjələ]	n. 公式，方程式
cross section	[krɒs' sekʃ(ə)n]	横截面
absorb	[əb'zɔːb]	v. 吸收（液体、气体等），使并入
blow out	[bləʊ aʊt]	吹除
form	[fɔːm]	n. 类型，种类，形式
plain	[pleɪn]	adj. 平的，朴素的，清楚的
split	[splɪt]	adj. 劈开的，分离的
slot	[slɒt]	n. 窄缝，扁口
shaft	[ʃɒft]	n. 轴，竖井，井筒

retract	[rɪ'trækt]	v.	收回，缩回，撤销
cling	[klɪŋ]	v.	抓紧，紧握，紧抱
curvature	['kɜːvətʃə(r)]	n.	弯曲，曲度，曲率
gap	[gæp]	n.	缺口，差距，间隙
transport	['trænspɔːt]	n.	运输，交通运输系统
retreat	[rɪ'triːt]	v.	退却，撤退，离开，离去
particularly	[pə'tɪkjələli]	adv.	尤其，特别
triple	['trɪpl]	adj.	三部分的，三人的，三倍的
vibrate	[vaɪ'breɪt]	v.	振动，(使)颤动，摆动
percentage	[pə'sentɪdʒ]	n.	百分率，百分比
suitable	['suːtəbl]	adj.	合适的，适宜的
applicable	[ə'plɪkəbl]	adj.	可应用的，适用的，合适的
procedure	[prə'siːdʒə(r)]	n.	程序
contrast	['kɒntrɒst]	n.	明显的差异，对比，对照
actuator	['æktjʊeɪtə]	n.	作动筒
adopt	[ə'dɒpt]	v.	采用(某方法)，收养，领养
frustrate	[frʌ'streɪt]	v.	使懊丧，使懊恼
eliminate	[ɪ'lɪmɪneɪt]	v.	排除，清除，消除
vortex	['vɔːteks]	n.	涡流，涡旋
rib	[rɪb]	n.	翼肋，肋骨，排骨
fix	[fɪks]	v.	修理，安装，使固定
automatic	[ˌɔːtə'mætɪk]	adj.	自动的，无意识的
sliding	['slaɪdɪŋ]	v.	(使)滑行，滑动
mechanism	['mekənɪzəm]	n.	机械装置，机件，方法
controllable	[kən'trəʊləbl]	adj.	可控制的
handle	['hændl]	n.	手柄，柄，把手
prior to	['praɪə(r) tuː]		在前，居先
deploy	[dɪ'plɔɪ]	v.	(使)张开，部署，调度(军队或武器)
expansion	[ɪk'spænʃn]	n.	膨胀，扩张，扩展，扩大
eventually	[ɪ'ventʃuəli]	adv.	最后，终于
suction	['sʌkʃn]	n.	吸，抽吸，吸出
vortex generator	['vɔːteks 'dʒenəreɪtə(r)]		涡流发生器
shock wave	[ʃɒk weɪv]		激波

Q&A

The following questions are for you to answer to assess the learning outcomes.

(1) Retelling the lift formula.

(2) What is the high-lift principle of the high lift device?

(3) What are the ways to increase the lift of an airplane?

(4) What devices can increase the lift of an aircraft?

(5) Which trailing flap has the best lift effect?

(6) Please briefly describe the airflow changes before and after the deploy of the leading edge slat.

(7) Please briefly describe the high-lift principle of vortex generator.

(8) Please talk about the influence on the drag when applying each high-lift device in combination with the lift formula and the resistance formula.

飞行稳定性
Flight Stability

Contents

1) Stability

2) Lateral (pitching) stability

3) Longitudinal (rolling) and vertical (yawing) stability

学习内容

1）稳定性

2）横向（俯仰）稳定性

3）纵向（滚转）和垂直（偏航）稳定性

任务 1 稳定性
Task 1 Stability

 Contents

1) Stability

2) Flight stability

3) Lateral stability

4) Longitudinal stability

5) Vertical stability

Learning Outcomes

1) Master the concept and classification of stability

2) Master the concept of aircraft stability

3) Cultivate professional qualities of rigor, carefulness, and ability to express, coordinate, and communicate effectively

 任务内容

　　1）稳定性

　　2）飞行稳定性

　　3）俯仰稳定性

　　4）滚转稳定性

　　5）偏航稳定性

 任务目标

　　1）掌握稳定性的概念和分类

　　2）掌握飞机稳定性的含义

　　3）培养严谨、细心的职业素养，以及有效表达、协调和沟通的能力

Learning Guide

The stability of an aircraft is an important parameter in aircraft design that measures flight quality, indicating whether the aircraft has the ability to return to its original state after being disturbed. If the aircraft is able to return to its initial state without any control by the pilot after being disturbed (such as gusts), it is considered stable; otherwise, it is considered unstable.

The concepts of stability and instability can be vividly explained. For example, if we place a small ball on the crest of a wavy surface and gently push it, it will leave the crest and fall into the trough. We refer to the state where the ball is at the crest as an unstable state. On the contrary, if we place the ball in the trough and gently push it, the ball can still return to the bottom after a period of oscillation, and we call the ball in a stable state.

课文

1. Stability

1. 稳定性

An object in balance is disturbed and then deviates from the balanced position. If the object can automatically return to the original balanced position, it can be said that, the balanced state of the object is steady, and that the balanced state of the object owns stability.

228

As shown in Fig. 7–1, the state of the first object (on the left) is stable, the state of the second object (middle) is neutrally stable, and the state of the third object (on the right) is not stable.

处于平衡状态的物体受到扰动，然后偏离平衡位置。物体能否自动返回到原始平衡位置取决于物体的平衡状态，即物体的平衡状态是否具有稳定性。

如图 7-1 所示，第一个物体（左边）的状态是稳定的，第二个物体（中间）的状态为中立稳定，第三个物体（右边）的状态为不稳定。

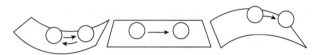

Fig. 7–1　Different states with different balanced state stability

图 7-1　不同的状态具有不同的状态稳定性

The stability of the balanced state can be divided into static stability and dynamic stability.

平衡状态的稳定性可分为静态稳定性和动态稳定性。

1) Static Stability

1）静态稳定性

Static stability is the tendency of an object to return to its original balanced state when the external disturbance disappears.

静态稳定性是当外部扰动消失时，物体具有返回其原始平衡状态的趋势。

2) Dynamic Stability

2）动态稳定性

Dynamic stability refers to the dynamic process of an object returning to its original balanced state when the external disturbance disappears (Fig. 7–2).

动态稳定性是指当外部扰动消失后，物体恢复到其原始平衡状态的动态过程（图 7-2）。

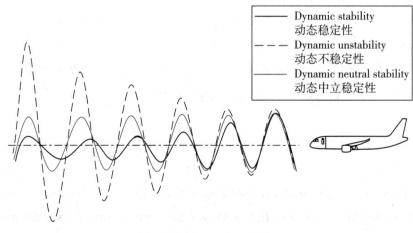

Fig. 7–2　Dynamic stability

图 7-2　动态稳定性

静态稳定性是平衡状态具有稳定性的必要非充分条件。只有具有动态稳定性的平衡状态才是真正稳定的。

Static stability is a necessary condition for the stability of balanced state, but it is not sufficient. Only the balanced state with dynamic stability is truly stable.

2. Flight Stability

2. 飞行稳定性

There are various disturbances in flight, such as gust that changes the angle of attack and flight speed, little deflection of the control surface caused by airflow.

The aerodynamic force and moment of the aircraft would also change with the disturbance. Then, the original balanced state of the aircraft would be changed during the process.

When the disturbance disappears, if the aircraft can automatically return to the original balanced state without the control of the pilot, we say the aircraft has stability, otherwise it will be unstable or neutrally stable.

The aircraft must have a certain degree of stability, which is very important for flight safety.

飞行中存在各种扰动，如改变迎角和飞行速度的阵风，以及气流引起的操纵面小偏转。

在扰动情况下，飞机的气动力和力矩也会发生变化，进而改变飞机的原始平衡状态。

当扰动消失后，如果飞机能够在没有飞行员控制的情况下自动恢复到原始平衡状态，飞机的飞行状态就是稳定的，具有稳定性；否则飞机就是不稳定的，或者是中立稳定的。

飞机必须具有一定程度的稳定性，这对飞行安全来说是非常重要的。

1) Lateral Stability

1）俯仰稳定性

When the aircraft is disturbed, it pitches around the lateral axis, and the angle of attack of the aircraft increases or decreases. If the aircraft could automatically return to the original flight state without pilot control, we say that the aircraft owns lateral stability, also known as lateral stability, that is, the stability around the lateral axis.

当飞机受到扰动时，它会绕着横轴进行俯仰，导致飞机的迎角增大或减小。飞机在没有飞行员控制的情况下自动返回原始飞行状态的能力称为俯仰稳定性，即飞机绕横轴的稳定性。

2) Longitudinal Stability

2）滚转稳定性

When the aircraft is disturbed, it rolls around the longitudinal axis. If the aircraft could automatically return to the original flight state without pilot control, we say the aircraft owns longitudinal stability, also known as rolling stability, that is, the stability around the longitudinal axis.

当飞机受到扰动时，它会绕纵轴进行滚转。飞机在没有飞行员控制的情况下自动返回原始飞行状态的能力称为滚转稳定性，即飞机绕纵轴的稳定性。

3) Vertical Stability

3）偏航稳定性

When the aircraft is disturbed, it yaws around the vertical axis. If the aircraft could automatically return to the original flight state without pilot control, we say the aircraft owns vertical stability, also known as yaw stability, that is, the stability around the vertical axis.

当飞机受到扰动时，它会绕垂直轴偏航。飞机在没有飞行员控制的情况下自动返回原始飞行状态的能力称为偏航稳定性，即飞机绕垂直轴的稳定性。

Stability (1)   Stability (2)   Stability (3)

 New Words

disturb	[dɪ'stɜːb]	vt.	打扰；扰动；妨碍；搅乱
automatically	[ˌɔːtə'mætɪkli]	adv.	自动地；机械地；无意识地
state	[steɪt]	n.	状态；状况；情况
stability	[stə'bɪləti]	n.	稳定性；稳定
stable	['steɪbl]	adj.	稳固的；牢固的
neutrally	['njuːtrəli]	adv.	中立地；保持中立地；不露倾向地
tendency	['tendənsi]	n.	趋势；倾向；趋向
disappear	[ˌdɪsə'pɪə(r)]	vi.	消失；不见；不复存在
process	[prə'ses]	n.	过程；进程
necessary	['nesəsəri]	adj.	必需的；必要的；必然的
sufficient	[sə'fɪʃnt]	adj.	足够的；充足的
disturbance	[dɪs'tɜːbəns]	n.	打扰，扰动，妨碍
unstable	[ʌn'steɪbl]	adj.	不稳定的；易变的；变化莫测的
degree	[dɪ'griː]	n.	度；程度；度数

 Q&A

The following questions are for you to answer to assess the learning outcomes.

(1) Describe the definition of stability.

(2) Describe the definition of static stability.

(3) Describe the definition of dynamic stability.

(4) Describe the definition of vertical stability of aircraft.

(5) Describe the definition of lateral stability of aircraft.

(6) Describe the definition of longitudinal stability of aircraft.

 Extended Reading

Stability

What do we mean by the stability of an aircraft? Fundamentally, we have to discern between the stability of the aircraft to external impetus, with and without the pilot responding to the perturbation. Here we will limit ourselves to the inherent stability of the aircraft. Hence the aircraft is said to be stable if it returns back to its original equilibrium state after a small perturbing displacement, without the pilot intervening. Thus, the aircraft's response arises purely from the inherent design. At level flight, we tend to refer to this as static stability. In effect the airplane is statically stable when it returns to the original steady flight condition after a small disturbance; statically unstable when it continues to move away from the original steady flight condition upon a disturbance; And neutrally stable when it remains steady in a new condition upon a disturbance. The more pernicious type of stability is dynamic stability. The airplane may converge continuously back to the original steady flight state; It may over-correct and then converge to the original configuration in a oscillatory manner; Or it can diverge completely and behave uncontrollably, in which case the pilot is well-advised to intervene. Static instability naturally implies dynamic instability, but static stability does not generally guarantee dynamic stability.

任务 2　俯仰稳定性
Task 2　Lateral Stability

 Contents

1) Lateral stability definition

2) Lateral balance

3) Lateral trim

4) Aircraft focus

5) Lateral static stability

6) Lateral static stability overmeasure

7) Lateral dynamic stability

8) Lateral dynamic motion modes

Learning Outcomes

1) Master the conditions and requirements of aircraft lateral stability

2) Master the forces of the aircraft in the process of lateral disturbance

3) Understand the motion mode in the process of aircraft lateral disturbance

4) Analyze the dynamics and flight quality problems in flight by using the lateral stability theory of the aircraft

5) Cultivate professional qualities of rigor, carefulness, and ability to express, coordinate, and communicate effectively

任务内容

1）俯仰稳定性定义

2）俯仰平衡

3）俯仰配平

4）飞机的焦点

5）俯仰静稳定性

6）俯仰静稳定裕度

7）俯仰动稳定性

8）飞机的俯仰扰动运动模式

任务目标

1）掌握飞机俯仰稳定性的条件和要求

2）掌握飞机在俯仰扰动过程中的受力情况

3）了解飞机在俯仰扰动过程中的运动模式

4）能够运用飞机的俯仰稳定性分析飞行中的动力学和飞行品质问题

5）培养严谨、细心的职业素养，以及有效表达、协调和沟通的能力

Learning Guide

The lateral stability of an aircraft refers to the characteristic of the aircraft being subjected to small disturbances during flight, resulting in the disruption of its lateral/ (pitching) balance. After the disturbance disappears, the aircraft automatically tends to return to its original equilibrium state.

 课文

1. Lateral Stability Definition
1. 飞机俯仰稳定性的定义

After the aircraft is disturbed, the process of returning to the original flight attitude is determined by the interaction of the lateral (pitching) static stability moment, lateral (pitching) damping moment and inertia moment of the aircraft.

飞机受到扰动后，返回原始飞行状态的过程由飞机的横向（俯仰）静稳定力矩、横向（俯仰）阻尼力矩和惯性力矩的相互作用决定。

1) Lateral Static Stability Moment
1）俯仰静稳定力矩

For the aircraft with lateral static stability in the balanced state, when the aircraft is disturbed and the angle of attack changes, the lateral (pitching) moment caused by the change of the angle of attack is the moment that makes the aircraft return to the original flight attitude, which is called the lateral static stability moment.

对于平衡状态下具有俯仰静稳定性的飞机，当飞机受到扰动，迎角发生变化时，由迎角变化引起的横向（俯仰）力矩是使飞机返回原始飞行姿态的力矩，称为俯仰静稳定力矩。

2) Lateral Damping Moment
2）俯仰阻尼力矩

In the process of pitching back to the original flight attitude, the aerodynamic force generated on the control surfaces forms a moment to prevent the lateral movement of the aircraft, which is called lateral damping moment (Fig. 7-3).

(1) The lift increment on the aircraft produces additional lateral moment.

(2) When the aircraft pitches up, the angle of attack in front of the CG decreases, and the lift increment acts downward. The angle of attack increases behind the CG, the lift increment acts upward. The moment formed by the distributed lift increment on the CG is a nose down moment, which prevents the aircraft from nosing up. This is the lateral damping moment.

(3) The horizontal tail is the farthest aerodynamic component from the CG of the aircraft. Therefore, the lateral damping moment is mainly generated by the horizontal tail.

在俯仰回到原始飞行姿态的过程中，操纵面上产生的气动力形成的力矩是阻止飞机的俯仰运动的，称为俯仰阻尼力矩（图7-3）。

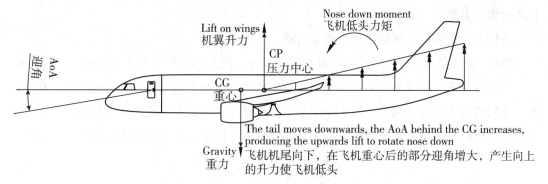

Fig. 7-3 Generation of the lateral damping moment

图 7-3 俯仰阻尼力矩的产生

（1）飞机的气动升力增量产生额外的俯仰力矩。

（2）当飞机俯仰时，重心前方的迎角减小，由此产生的升力增量方向向下；在重心后方的部位迎角增大，产生的升力增量方向向上。由重心上的分布升力增量形成的力矩让飞机低头，阻止飞机抬头。这就是俯仰阻尼力矩。

（3）水平尾翼是离飞机重心最远的气动部件。因此，俯仰阻尼力矩主要由水平尾翼产生。

3) Inertia Moment

3）惯性力矩

In the lateral motion of the aircraft returning to the original flight attitude, due to the angular acceleration of the fuselage rotating around the center of gravity, the rotational inertia makes the aircraft to maintain the original motion state, which is called inertia moment.

在飞机返回原始飞行姿态的俯仰运动中，由于机身绕重心旋转的角加速度，转动惯性使飞机保持原始的运动状态，这称为惯性力矩。

2. Lateral Balance

2. 飞机的俯仰平衡

The aircraft is composed of wings, fuselage, tail and power plant. The aerodynamic forces and engine thrust on each of these component will produce lateral moment to the aircraft.

The lateral moment makes the aircraft pitch around the lateral axis.

The lateral moment (Fig. 7-4) of the whole aircraft is the sum of the lateral moment generated on the wings, fuselage, tail and other components.

Generally, the CP of the wing is behind the CG of the aircraft. The aerodynamic lift on the wing acts upwards and produces a lateral moment on the aircraft that makes the nose down. The aerodynamic lift on the horizontal tail acts downward, producing the nose up lateral moment.

When the two moments counter each other, the aircraft maintains lateral balance state.

飞机由机翼、机身、尾翼和动力装置组成。这些部件上的气动力和发动机推力将对飞机产生俯仰力矩。

俯仰力矩使飞机绕横轴线产生俯仰运动。

整个飞机的俯仰力矩（图7-4）是机翼、机身、尾部和其他部件产生的俯仰力矩之和。

一般情况下，机翼的压力中心在飞机重心之后。机翼上的气动升力方向向上，在飞机上产生的俯仰力矩使飞机低头。水平尾翼上的气动升力的方向向下，在飞机上产生的俯仰力矩使飞机抬头。当这两个力矩相互达到平衡时，飞机保持俯仰平衡的状态。

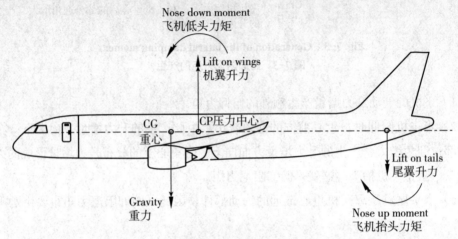

Fig. 7-4　Primary lateral moments of the aircraft
图7-4　飞机上主要的俯仰力矩

To make the horizontal tail produce the nose up moment, the installation angle of the horizontal tail generally is a negative value.

An important role of the horizontal tail of an aircraft is to maintain the lateral balance of the aircraft in steady straight flight at different speeds.

为了使水平尾翼产生抬头的力矩，水平尾翼的安装角度通常为负值。

飞机水平尾翼的一个重要作用是确保飞机在不同速度下稳定直线飞行时的俯仰平衡。

3. Lateral Trim

3. 俯仰配平

When an aircraft flies in a steady straight direction, different speed requires different angles of attack. With different angles of attack, the wing lift and the position of the CP are also different, which will produce different lateral moments to the CG of the aircraft. Therefore, it is necessary to change the deflection angle of the elevator (some aircraft can also change the trim angle of the horizontal stabilizer) to make the horizontal tail generate a balanced moment to maintain the lateral balance of the aircraft. This process is called the lateral trim of the aircraft.

当飞机做稳定直线飞行时，不同的速度需要不同的迎角。对于不同的迎角，机翼升力和压力中心的位置也不同，这将对飞机的重心产生不同的俯仰力矩。因此，有必要改变升降舵的偏转角（一些飞机还可以改变水平安定面的配平角），使水平尾翼产生平衡的力矩，以保持飞机的俯仰平衡。这个过程称为飞机的俯仰配平。

4. Aircraft Focus

4. 飞机的焦点

When the aircraft is disturbed, the AoA of the wing, fuselage, and horizontal tail will change, generating additional aerodynamic lift. The sum of these additional lift makes the increment of the lift of the whole aircraft. The acting point of the lift increment caused by the AoA change is called the focus of the aircraft (Fig. 7-5).

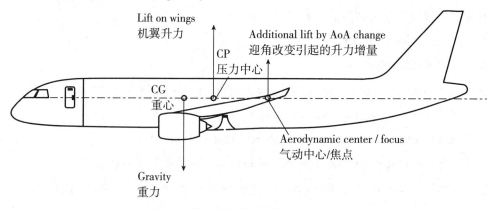

Fig. 7-5　Aircraft focus

图 7-5　飞机的焦点

Since the lift of the wing accounts for the main part of the lift of the whole aircraft, the position of the focus of the aircraft mainly depends on the focus of the wing. Although the lift on the horizontal tail is much smaller than that on the wing, its arm is distant after the CG of the aircraft, therefore, the focus of the aircraft is located after the wing focus.

The focus position of the wing remains basically unchanged, when the aircraft Mach number is less than the critical Mach number, the position of the wing focus is at about 25% of the wing chord. The position of the whole aircraft focus remains basically the same when the aircraft flies at low speed. The focus moves backward when the aircraft enters the supersonic flight speed.

The cargo, passengers, consumption of fuel, and the configuration of the aircraft will affect the position of the practical CG of the aircraft.

The position of the focus would also be affected by the deflection angle of elevators, trim angle of horizontal stabilizers, position of flaps, position of slats, position of landing gears, and installation clearance of the lateral control surfaces.

当飞机受到扰动时，机翼、机身和水平尾翼的迎角将发生变化，产生附加的气动升力。这些附加升力之和形成了整个飞机的气动升力增量。由迎角变化引起的升力增量的作用点称为飞机的焦点（图7-5）。

由于机翼升力占飞机升力的主要部分，因此飞机的焦点位置主要取决于机翼的焦点位置。虽然水平尾翼上的升力比机翼上升力的小得多，但它的作用力臂位于重心之后较远的位置，因此，飞机的焦点位置在尾翼的作用下，位于机翼焦点之后。

当飞机马赫数小于临界马赫数时，机翼的焦点位置基本保持不变，约在机翼弦线的25%处。当飞机低速飞行时，整个飞机焦点的位置基本保持不变。超声速飞行时，飞机的焦点明显后移。

货物、乘客、燃油消耗以及飞机的配置等，都将影响飞机的实际重心。

焦点位置也会受到升降舵偏转角、水平安定面配平角、襟翼位置、缝翼位置、起落架位置和俯仰操纵面的安装间隙的影响。

5. Lateral Static Stability

5. 飞机的俯仰静稳定性

When the flight AoA is less than the critical angle of attack, the lateral static stability of the aircraft only depends on the position relation of the focus and the aircraft's CG.

(1) The horizontal tail provides lateral static stability for the aircraft.

(2) Lateral static stability: the focus of the aircraft is behind the CG (Fig. 7–6).

(3) Lateral static instability: the focus of the aircraft is in front of the CG (Fig. 7–7).

(4) Lateral neutral static stability: the focus and CG of the aircraft are in the same position.

当飞行迎角小于临界迎角时，飞机的俯仰静稳定性仅与整个飞机的焦点和重心的相对位置（位置关系）有关。

（1）水平尾翼为飞机提供俯仰静稳定性。

（2）俯仰静稳定性：飞机的焦点位于重心之后（图7-6）。

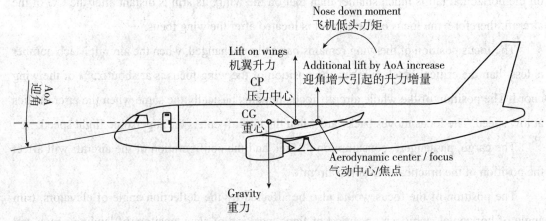

Fig. 7–6　Lateral static stability

图 7–6　俯仰静稳定性

（3）俯仰静不稳定性：飞机的焦点位于重心之前（图 7-7）。

（4）俯仰中立静稳定性：飞机的焦点和重心重合。

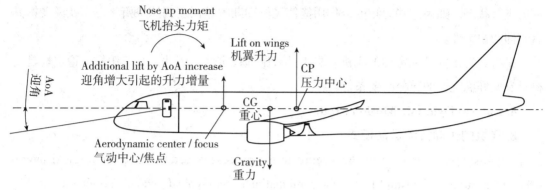

Nose up moment
飞机抬头力矩

Lift on wings
机翼升力

Additional lift by AoA increase
迎角增大引起的升力增量

CP
压力中心

AoA
迎角

CG
重心

Aerodynamic center / focus
气动中心/焦点

Gravity
重力

Fig. 7-7　Lateral static instability

图 7-7　俯仰静不稳定性

6. Lateral Static Stability Overmeasure

6. 俯仰静稳定裕度

The distance between the focus and the aircraft's CG is called the lateral static stability overmeasure.

To ensure a certain lateral static stability of the aircraft, the focus of the aircraft is required to be at a certain distance behind the CG. Different aircraft have different requirements for the distance.

The lateral static stability is reduced during stick free flight, because the focus moves forward.

焦点和飞机重心之间的距离称为俯仰静稳定裕度。

为了确保飞机具有一定的俯仰静稳定性，要求飞机的焦点位于重心后方一定距离处。不同的飞机对重心到焦点的距离有不同的要求。

在松杆飞行期间，俯仰静态稳定性降低，即焦点位置向前移动。

7. Lateral Dynamic Stability

7. 飞机的俯仰动稳定性

After the disturbance stops, there is the combination of static stability moment, lateral damping moment, inertia moment acting on the aircraft to make it return to the original flight attitude. If the combined action of these moments can make a gradual converge on the balanced state, and the aircraft finally returns to its original flight attitude, we can say that the aircraft has lateral dynamic stability.

Only aircraft with lateral dynamic stability can truly have lateral stability. Therefore, having lateral dynamic stability is a sufficient and necessary condition for an aircraft to have lateral stability.

扰动消失后，作用在飞机上的静稳定力矩、俯仰阻尼力矩和惯性力矩将让飞机返回之前的飞行状态。如果这些力矩的合作用能使飞机逐渐收敛回复到原平衡状态，就说飞机具有俯仰动稳定性。

只有具有俯仰动稳定性的飞机才真正具有俯仰稳定性。因此，具有俯仰动稳定性是飞机具有俯仰稳定性的充分必要条件。

8. Lateral Dynamic Motion Modes

8. 飞机的俯仰扰动运动模式

The disturbance motion of the aircraft in the process of returning to the original lateral balanced attitude can be simplified as the combination of two typical periodic motion modes.

飞机在返回初始俯仰平衡状态过程中的扰动运动可以简化为两种典型周期运动模式的叠加。

1) The Short Term Motion Mode

1）短周期运动模式

It mainly occurs in the initial stage after the interference disappears (Fig. 7-8).

The disturbance motion of the aircraft is mainly the swing process of the aircraft around its CG, which causes the angle of attack and angular velocity change rapidly periodically, while the flight speed remains basically unchanged. The aerodynamic lift on the aircraft produces a large lateral damping moment that is opposite to the direction of the aircraft's rotation, which makes the lateral swing of the aircraft decrease quickly.

Generally, this short term swing of the aircraft disappears in the first few seconds.

这种模式主要发生在扰动消失后的开始阶段（图 7-8）。

飞机的扰动运动主要是飞机绕其重心的摆动过程，这使飞机的迎角和角速度周期性地快速变化，然而飞行速度基本保持不变。飞机上的升力产生与飞机俯仰方向相反的俯仰阻尼力矩，使飞机的俯仰摆动迅速减小。

通常，飞机的这种短周期的摆动振荡基本上在前几秒结束。

Fig. 7-8　Short term motion mode

图 7-8　短周期运动模态

2) The Long Term Motion Mode

2）长周期运动模式

It mainly occurs in the following stage of disturbance motion (Fig. 7-9).

This motion mode of the aircraft is mainly the swing process of the CG of the aircraft, which is the periodic and slow changes of flight speed and path angle. The angle of attack of the aircraft returns to the original angle of attack and keeps unchanged. The track of the aircraft is curved.

This swing process decays very slowly and forms a long term motion mode.

Lateral Stability (1)

这种模式主要发生在扰动运动的后期（图7-9）。

飞机的扰动运动主要是飞机重心的振荡过程，表现为飞行速度和航迹角的周期性缓慢变化。飞机的迎角基本上恢复到原始迎角并保持不变。飞机的飞行轨迹是弯曲的。

该振荡过程衰减非常缓慢，形成长周期运动模式。

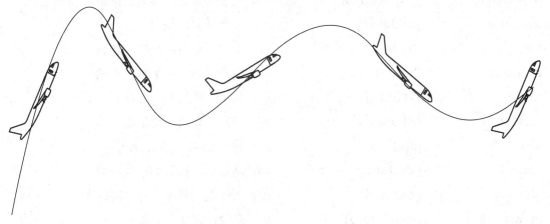

Fig. 7-9　Long term motion mode
图 7-9　长周期运动模态

Lateral Stability (2)　Lateral Stability (3)　Lateral Stability (4)　Lateral Stability (5)　Lateral Stability (6)

Lateral Stability (7)　Lateral Stability (8)　Lateral Stability (9)　Lateral Stability (10)　Lateral Stability (11)

New Words

compose	[kəm'pəʊz]	v.	组成；作曲；构成
power plant	['paʊə plɒnt]		动力装置；发电厂
sum	[sʌm]	n.	总和；和；金额
generate	['dʒenəreɪt]	vt.	生成；产生；引起
pressure center	['preʃə(r) 'sentə(r)]		压力中心
center of gravity	['sentə(r) əv 'grævəti]		重心
counter	['kaʊntə(r)]	adv.	逆向地；相反地；反对地
horizontal	[ˌhɒrɪ'zɒntl]	adj.	水平的；与地面平行的
installation	[ˌɪnstə'leɪʃn]	n.	安装；设置；安装的设备
deflection	[dɪ'flekʃn]	n.	突然转向，偏斜
trim	[trɪm]	vt.	调整；修剪；修整；切去
increment	['ɪŋkrəmənt]	n.	定期的加薪；增量；增加
focus	['fəʊkəs]	v.	集中（注意力、精力等于）
account	[ə'kaʊnt]	n.	账户；账目；赊销账
arm	[ɒm]	n.	臂；手臂；上肢；袖子
distant	['dɪstənt]	adj.	遥远的；远处的；久远的
basically	['beɪsɪkli]	adv.	大体上；基本上；总的说来
overmeasure	[əʊvə'meʒə]	n.	剩余；裕度；容差
stick	[stɪk]	v.	粘贴；将……刺入（或插入）
affect	[ə'fekt]	v.	影响；侵袭；使感染
clearance	['klɪərəns]	n.	清除；排除；清理
damping	['dæmpɪŋ]	v.	抑制；弄潮；使潮湿
inertia	[ɪ'nɜːʃə]	n.	缺乏活力；惰性
additional	[ə'dɪʃənl]	adj.	附加的；额外的；外加的
periodic	[ˌpɪəri'ɒdɪk]	adj.	周期性；周期的
typical	['tɪpɪkl]	adj.	典型的；有代表性的
mode	[məʊd]	n.	模式；方式；风格；样式
swing	[swɪŋ]	v.	摆动；（使）摇摆；摇荡
rapidly	['ræpɪdlɪ]	adv.	迅速地；迅速；高速
decay	[dɪ'keɪ]	v.	衰退；（使）腐烂

242

Q&A

The following questions are for you to answer to assess the learning outcomes.

(1) Describe the positional relationship between the aircraft center of pressure and the aircraft center of gravity.

(2) How does the lateral moment of an airplane make it move?

(3) Why is the installation angle of aircraft horizontal tail less than zero?

(4) What conditions should an aircraft meet to maintain lateral static stability?

 Extended Reading

Longitudinal Control and Static Stability[①]

Longitudinal stability and pitch control of an aircraft are the most basic properties which concern a pilot. Most of the time, a pilot wants to hold an aircraft at a constant incidence, and does so by moving a control surface to the "right" position for moment equilibrium. In order to change the state of flight, the pilot moves the control surface to some other positions to impose a finite moment on the aircraft, and force it to rotate. The basic instrument for analysis of the aircraft is thus a moment equation, derived from a free body diagram. Fig. 7–10 shows some tailless aircraft.

Vulcan:delta wing
三角翼

Saab Gripen:close–coupled canard
耦合机翼

Pegasus Quantum15–912:flex–wing
软性机翼

Fig. 7–10　Some tailless aircraft
图 7–10　一些无尾式飞机的机翼平面

Fig. 7–11 shows the control surfaces on a tailless aircraft. The rudder operates as on a conventional layout, but elevators and ailerons are combined into "elevons" which operate differently for roll control, and together in pitch.

---

① 文中为了表述不引起混淆，将操纵性和稳定性在"俯仰"方向表示为"Lateral"，即以机动的坐标轴对姿态进行定义。参考文献中"俯仰"方向另表示为"Longitudinal"，请读者注意区分。

Rudder
方向舵

Elevons
升降舵副翼

Fig. 7-11　Control surfaces for tailless aircraft

图 7-11　无尾式飞机操纵面

The variables relevant to analyzing a tailless aircraft are shown in Fig. 7-12. Clearly, there is no tailplane contribution to include in calculating the lateral moment, but there is a complication because elevon deflection generates a change in lift coefficient and a change in lateral moment. The coupling of these two effects can make tailless aeroplanes quite challenging to control, especially on landing.

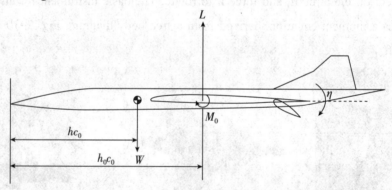

Fig. 7-12　Tailless aircraft

图 7-12　无尾式飞机

The lift coefficient for a tailless aircraft looks a bit like the corresponding expression for a tailplane, because there is a control deflection to include, with no need to consider the stick free case because such aeroplanes usually have powered controls.

任务 3 飞机的滚转和偏航稳定性
Task 3　Longitudinal and Vertical Stability

Contents

1) Longitudinal and vertical stability

2) Longitudinal static stability

3) Vertical static stability

4) Longitudinal and vertical dynamic stability

5) Interfering moment

6) The motion modes of roll and yaw

Learning Outcomes

1) Master the conditions and requirements of aircraft longitudinal and vertical stability

2) Master the forces of the aircraft of longitudinal and vertical disturbance

3) Understand the motion mode of longitudinal and vertical disturbance

4) Analyze the dynamics and flight quality problems in flight by using the longitudinal and vertical stability theory of the aircraft

5) Cultivate professional qualities of rigor, carefulness, and ability to express, coordinate, and communicate effectively

任务内容

1）滚转和偏航稳定性

2）滚转静稳定性

3）偏航静稳定性

4）滚转和偏航动稳定性

5）交叉力矩

6）滚转和偏航扰动的运动模式

任务目标

1）掌握飞机滚转和偏航稳定性的条件与要求

2）掌握飞机在滚转和偏航扰动过程中的受力情况

3）了解飞机滚转和偏航扰动过程中的运动模态

4）能够运用飞机的滚转和偏航稳定性分析飞行中的动力学和飞行品质问题

5）培养严谨、细心的职业素养，以及有效表达、协调和沟通的能力

 Learning Guide

The stability of an aircraft can be generated by its own aerodynamic performance or by advanced autopilots. Regardless of which generation method, fundamentally, it generates a stabilizing force that restores the aircraft to a stable state when it deviates from its original state.

 课文

1. Longitudinal and Vertical Stability

1. 飞机的滚转和偏航稳定性

There are three degrees of freedom in the lateral motion of the aircraft: roll, yaw, and side sliding. The roll or yaw of the aircraft will cause the aircraft to side slip (Fig. 7-13). The longitudinal and vertical stability are not independent, but interact with each other. For aircraft with longitudinal and vertical static stability, the longitudinal and vertical moment caused by the side slip are the moment that makes the aircraft to return to the original flight attitude after the disturbance disappears, which is the static stability moment. The longitudinal moment caused by longitudinal and the vertical moment caused by yaw are the damping moment generated by aerodynamic forces in the disturbed motion. To ensure the stability, it is necessary to maintain an appropriate proportion between the longitudinal static stability and the vertical static stability of the aircraft.

飞机的横侧向运动有三个自由度：滚转、偏航和侧滑。飞机的滚转或偏航都将导致飞机的侧滑（图7-13）。滚转和偏航稳定性不是独立的，而是相互影响的。对于具有滚转和偏航静稳定性的飞机，侧滑引起的滚转和偏航力矩是扰动消失后使飞机返回原始飞行状态的力矩，即静稳定性力矩。由滚转引起的滚转力矩和由偏航引起的偏航力矩是扰动运动中的气动力产生的阻尼力矩。为了确保稳定性，有必要在飞机的滚转静稳定性和偏航静稳定性之间保持适当的比例。

1) Longitudinal Static Stability

1）飞机的滚转静稳定性

The aircraft is disturbed and rotates around the longitudinal axis of its body, resulting in a rolling and side slipping. If the longitudinal moment caused by the side slipping is opposite to the direction of the aircraft's rolling, the aircraft has longitudinal static stability.

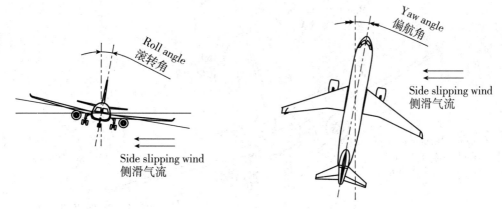

Fig. 7-13　Side slip caused by roll and yaw

图 7-13　滚转和偏航引起的侧滑

The main factors that affect the longitudinal static stability of aircraft are the wing's anhedral and sweep back angle.

The longitudinal static stability of the aircraft is mainly provided by the anhedral configuration of the wing. The anhedral angle of the wing not only provides the longitudinal static stability for the aircraft, but also makes it possible to quantitatively adjust the longitudinal static stability by changing the angle of the anhedral (Fig. 7-14).

The sweep back angle of the wing also provides longitudinal static stability for the aircraft (Fig. 7-15).

飞机受到扰动并绕机身纵轴滚转，导致滚动和侧滑。如果侧滑引起的滚转力矩与飞机滚转方向相反，则飞机具有滚转静稳定性。

影响飞机滚转静稳定性的主要因素是机翼的上反角和后掠角。

飞机的滚转静稳定性主要由机翼的上反角构型提供。机翼的上反角不仅为飞机提供了滚转静稳定性，还可以通过改变上反角的角度来定量地调整滚转静稳定性的大小（图 7-14）。

机翼的后掠角也为飞机提供了滚转静稳定性（图 7-15 ）。

The fin also affects the longitudinal static stability of the aircraft. The fin that is above the longitudinal axis of the aircraft increases the longitudinal static stability.

The relative position of the wing and the fuselage also affects the longitudinal static stability. The high wing plays a role of longitudinal static stability, and the lower wing plays a role of longitudinal static instability.

垂尾也会影响飞机的滚转静稳定性。位于飞机纵轴上方的垂尾会增加滚转静稳定性。

机翼和机身的相对位置也会影响滚转静稳定性。上单翼起滚转静稳定作用，下单翼起滚转静不稳定作用。

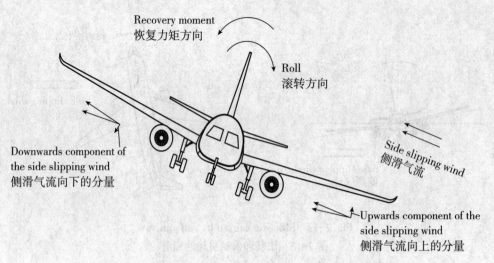

Recovery moment
恢复力矩方向

Roll
滚转方向

Downwards component of
the side slipping wind
侧滑气流向下的分量

Side slipping wind
侧滑气流

Upwards component of the
side slipping wind
侧滑气流向上的分量

Fig. 7-14 Longitudinal static stability with anhedral angle
图 7-14   由上反角提供的静稳定性

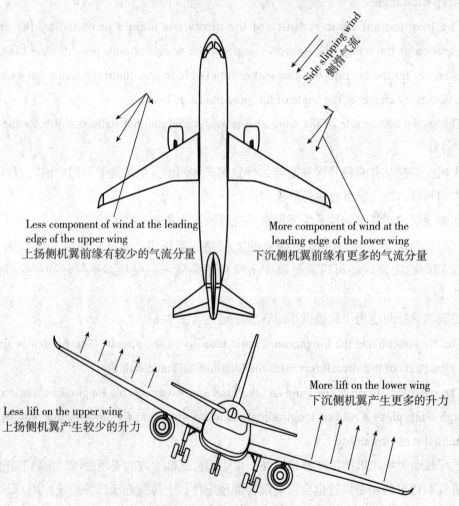

Side slipping wind
侧滑气流

Less component of wind at the leading
edge of the upper wing
上扬侧机翼前缘有较少的气流分量

More component of wind at the
leading edge of the lower wing
下沉侧机翼前缘有更多的气流分量

Less lift on the upper wing
上扬侧机翼产生较少的升力

More lift on the lower wing
下沉侧机翼产生更多的升力

Fig. 7-15   Longitudinal static stability with sweep back angle
图 7-15   由后掠角提供的滚转静稳定性

2) Vertical Static Stability
2）飞机的偏航静稳定性

When the aircraft rotates around the vertical axis due to disturbance and generates a side slipping, if the vertical moment caused by the side slipping always tries to align the aircraft with the relative airflow and eliminate the side slipping, the aircraft will have vertical static stability.

The main factors affecting the vertical static stability are the area of the fin and its length of the arm to the CG of the aircraft.

The vertical static stability provided by the fin always makes the nose of the aircraft to align with the relative wind and eliminate the side slipping (Fig. 7-16).

当飞机因扰动绕垂直轴转动并产生侧滑时，如果侧滑引起的偏航力矩始终试图使飞机对准相对气流并消除侧滑，则飞机将具有偏航静稳定性。

影响偏航静稳定性的主要因素是垂尾的面积及其臂到飞机重心的长度（力臂）。

垂尾提供的偏航静稳定力矩始终使飞机机头对准相对气流，并消除侧滑（图 7-16）。

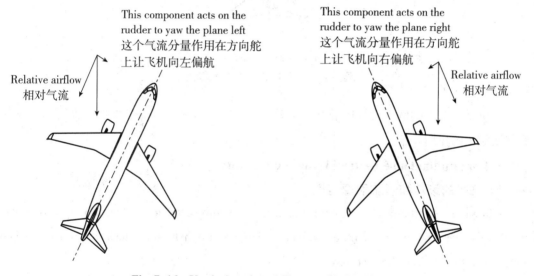

Fig. 7-16   Vertical static stability provided by the fin
图 7-16　由垂尾提供的静稳定性

The unbalanced drag on both sides of the sweep back wing will align the nose of the aircraft with the relative airflow to eliminate the side slipping (Fig. 7-17).

后掠机翼两侧的阻力不平衡情况，将使飞机机头与相对气流对齐，以消除侧滑（图 7-17）。

The side frontal area in front of the aircraft's CG plays a role of vertical instability. The side frontal area behind the CG plays a role of vertical stability.

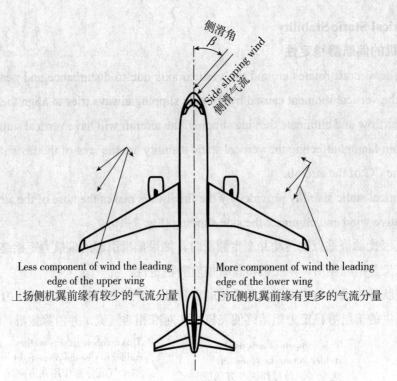

图中标注：

侧滑角 β

Side slipping wind 侧滑气流

Less component of wind the leading edge of the upper wing
上扬侧机翼前缘有较少的气流分量

More component of wind the leading edge of the lower wing
下沉侧机翼前缘有更多的气流分量

Fig. 7-17 Vertical static stability provided by the sweep back
图 7-17 由后掠翼提供的静稳定性

飞机重心前方的侧面迎风区域起着偏航静不稳定的作用。重心后面的侧面迎风区域起着偏航稳定的作用。

3) Longitudinal and Vertical Dynamic Stability

3）飞机的滚转和偏航动稳定性

After the disturbance disappears, the aircraft is subjected to the combination of static stability moment, inertia moment, aerodynamic damping moment and interfering moment when it returns to the original flight attitude.

The dynamic stability of rolling and yawing is the aerodynamic damping moment during rolling and yawing.

The longitudinal and vertical static stability moment is the moment recovery moment back to the original flight state due to the side slipping.

The moment of inertia is the moment generated by the inertia of the aircraft to keep its original state when the aircraft accelerates around the longitudinal and the vertical axis. The moment of inertia is influenced by factors such as the size, mass and distribution of the aircraft structure.

The aerodynamic damping moment is generated by the aerodynamic changes on the wing and fin during the rolling and yawing of the aircraft, which is opposite to the direction of the existing rolling and yawing. The wing plays a major role in the aerodynamic damping moment caused by rolling, and the fin plays a major role in the aerodynamic damping moment caused by yawing.

扰动消失后，飞机在返回原始飞行姿态的过程中受到静稳定力矩、惯性力矩、气动阻尼力矩和交叉力矩的共同作用。

飞机在偏航和滚转中的动稳定性主要由偏航和滚转过程中的气动阻尼力矩提供。

滚转和偏航静稳定力矩是由侧滑引起的恢复力矩。

惯性力矩是指当飞机绕纵轴和垂直轴加速时，由飞机的惯性产生的力矩。该力矩力图使飞机保持其原始状态。惯性矩的大小与飞机结构的尺寸、质量和分布等因素有关。

气动阻尼力矩由飞机滚转和偏航期间机翼和垂尾上的气动变化产生，与现有滚转和偏航运动方向相反。机翼在由滚转引起的气动阻尼力矩中起主要作用，垂尾在由偏航引起的气动力阻尼力矩中起主要作用。

4) Interfering Moment

4）交叉力矩

The maneuvers of rolling and yawing affect each other. The maneuver of rolling would cause yawing, and the maneuver of yawing would cause rolling.

When the aircraft rotates around the longitudinal axis and rolls to the right, the angle of attack of the left wing decreases and its drag decreases. As the angle of attack of the right wing increases, its drag increases. The unbalanced drag of the wings on both sides produces a vertical moment that rotates the nose to the right.

When the aircraft rolls to the right, the fin will also move to the right and down, causing the air flow through the fin to produce a rightward angle of attack. The aerodynamic forces on both sides of the fin are unbalanced, generating a left hand aerodynamic force, which also produces a vertical moment that deflects the nose to the right. This is the yawing caused by rolling.

When the aircraft yaw to the left around the vertical axis, the fin moves to the right, causing the air flow through the fin to produce an angle of attack to the right. The aerodynamic forces on both sides of the fin are unbalanced, resulting in forces pointing to the left. Since the its action point is at a certain distance to the longitudinal axis of the aircraft (arm), a longitudinal moment that makes the aircraft roll to the left is generated.

When the aircraft yaws to the left, the airspeed on the left wing decreases and its lift decreases. The airspeed on the right wing increases, and its lift increases. The unbalanced lift of the wings on both sides also produces a longitudinal moment that makes the aircraft roll to the left around the longitudinal axis.This is the longitudinal moment caused by yaw motion.

滚转和偏航是相互耦合的。交叉力矩是由滚转引起的偏航力矩和由偏航引起的滚转力矩。

当飞机绕纵轴旋转并向右滚转时，左翼迎角减小，阻力减小；右翼迎角增大，阻力增大。两侧机翼的不平衡阻力产生偏航力矩，使机头向右偏航。

当飞机向右滚转时，尾翼也会向右和向下移动，使气流通过尾翼产生向右的迎角。垂尾两侧的气动力不平衡，产生向左的气动力，这也会产生偏航力矩，使机头向右偏航。这就是由滚动引起的偏航。

当飞机绕立轴向左偏航时，尾翼相对于气流向右下运动，导致尾翼的气流产生向右的迎角。尾翼两侧的空气动力不平衡，合力方向指向飞机左侧。由于空气动力作用点与飞机纵轴之间有一定距离（力臂），因此产生了使飞机绕纵轴向左滚转的滚转力矩。

当飞机向左偏航时，左翼气流的相对速度减小，升力减小；右翼气流的相对速度增加，升力增加。两侧机翼的不平衡升力也会产生滚转力矩，使飞机绕纵轴向左滚转。这是由偏航运动引起的滚转力矩。

2. The Motion Modes of Rolling and Yawing

2. 飞机的滚转和偏航扰动的运动模式

There are 3 types of motion modes of rolling and yawing, when the aircraft returns back to its original state after the disturbance disappears.

当飞机在扰动消失后返回其原始状态时，有 3 种类型的滚转和偏航运动模式。

1) Rolling Convergent Mode

1）滚转收敛模式

The rolling convergent mode is a non-periodic mode, it decays rapidly.

The rolling angle and speed of the aircraft change in the rolling. The side slipping angle and yaw angle are very small and can be ignored. This is an approximate simple rolling around the longitudinal axis of the aircraft. Since the rolling inertia of the aircraft is small and the rolling damping moment is large, this rolling motion decays quickly.

Generally, aircraft can meet the stability requirements of this mode.

滚转收敛模式是一种非周期且快速衰减的运动模式。

飞机的滚转角和滚转速度在滚转过程中变化很快，而侧滑角和偏航角的变化很小，可以忽略不计。这种模态类似绕飞机纵轴的近似简单滚转运动。由于飞机的滚转惯性小，滚转阻尼力矩大，这种滚转运动衰减很快。

一般来说，飞机可以满足该模式的稳定性要求。

2) Spiral Mode

2）螺旋模式

Spiral mode is a non-periodic motion mode whose motion parameters change slowly (Fig. 7-18).

In the spiral mode motion, the side slipping angle is approximately zero, and the yaw angle is greater than the rolling angle, so the spiral mode motion is mainly a yaw motion with slight rolling. This unstable mode will appear when the vertical static stability of the aircraft is much greater than the rolling static stability.

The motion parameters of this mode change slowly, the pilots have enough time to correct, so it will not bring significant harm to flight safety.

螺旋模式是一种非周期运动模式，其运动参数变化缓慢（图 7-18）。

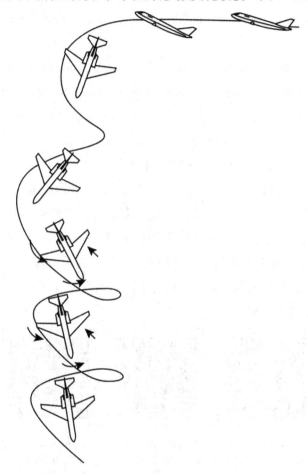

Fig. 7-18　Spiral mode
图 7-18　螺旋模态

在螺旋模式运动中，侧滑角近似为零，偏航角大于滚转角，因此螺旋模式运动主要是具有轻微滚转的偏航运动。当飞机的偏航静稳定性远大于滚转静稳定性时，将出现这种不稳定模式。

该模式的运动参数变化缓慢，飞行员有足够的时间进行校正，因此不会对飞行安全造成重大危害。

3) Dutch Roll Mode
3）荷兰滚模式

Dutch roll is a moderately damped longitudinal and vertical combination mode with high motion frequency.

In the dutch roll mode, the angles of side slipping, rolling, and yawing are almost the same, while the speed of rolling and yawing is smaller. All motion parameters change periodically and rapidly with time, and the aircraft rolls left and right, yaws left and right with side slipping, that is, dutch roll. This mode occurs when the longitudinal static stability is larger than the vertical static stability.

When the dutch roll occurs, due to its high frequency and short period, the aircraft will quickly shake left and right with gradually increasing amplitude. It is difficult for pilots to control this high frequency motion, so the dutch roll mode would seriously affect the flight safety. Therefore, the dutch roll must be positively damped and adjusted.

荷兰滚是一种具有高频的中度阻尼的滚转和偏航振动模式。

在荷兰滚模式下，侧滑、滚转和偏航角几乎相同，而滚转和偏航的速度较小。所有运动参数随时间周期性变化，飞机左右滚转，左右偏航，并伴随侧滑，即荷兰滚模态。当滚转静稳定性大于偏航静稳定性时，飞机容易发生荷兰滚。

当荷兰滚发生时，由于振荡频率高且周期短，飞机将以逐渐增大的幅度快速左右晃动。飞行员很难控制这种高频振荡，因此荷兰滚模式将严重影响飞行安全。因此，荷兰滚模式必须进行正向阻滞和调整。

Longitudinal and Vertical Stability (1) Longitudinal and Vertical Stability (2) Longitudinal and Vertical Stability (3) Longitudinal and Vertical Stability (4) Longitudinal and Vertical Stability (5)

Longitudinal and Vertical Stability (6) Longitudinal and Vertical Stability (7) Longitudinal and Vertical Stability (8) Longitudinal and Vertical Stability (9) Longitudinal and Vertical Stability (10)

New Words

roll	[rəʊl]	v.	滚转；（使）翻滚；滚动
yaw	[jɔː]	vi.	偏航
disturbance	[dɪˈstɜːbəns]	n.	骚乱；障碍；紊乱
anhedral	[ˈænhdrəl]	n.	（机翼的）正上反角
fin	[fɪn]	n.	垂尾；(鱼的) 鳍
sweep back	[swiːp bæk]		后掠翼
convergence	[kənˈvɜːdʒəns]	n.	汇聚；趋同
spiral	[ˈspaɪrəl]	n.	螺旋；盘旋；螺旋形的
dutch roll	[dʌtʃ rəul]		荷兰滚

 Q&A

The following questions are for you to answer to assess the learning outcomes.

(1) How does the longitudinal moment of an airplane make it move?

(2) Describe the definition of the longitudinal static stability.

(3) What parts of an aircraft can provide longitudinal static stability?

(4) How does the vertical moment of an airplane make it move?

(5) Describe the definition of the vertical static stability.

(6) What parts of an aircraft can provide vertical static stability?

(7) Describe the definition of dutch roll.

 Extended Reading

Slip and Side Slip

A slip is an aerodynamic state where an aircraft is moving somewhat sideways as well as forward relative to the oncoming airflow. In other words, for a conventional aircraft, the nose will be pointing in the opposite direction to the bank of the wing(s). The aircraft is not in coordinated flight and therefore is flying inefficiently.

1. Background

Flying in a slip is aerodynamically inefficient, since the lift to drag ratio is reduced. More drag is at play consuming energy but not producing lift. Inexperienced or inattentive pilots will often enter slips unintentionally during turns by failing to coordinate the aircraft with the rudder. Airplanes can readily enter into a slip climbing out from takeoff on a windy day. If left unchecked, climb performance will suffer. This is especially dangerous if there are nearby obstructions under the climb path and the aircraft is underpowered or heavily loaded.

A slip can also be a piloting maneuver where the pilot deliberately enters one type of slip or another. Slips are particularly useful in performing a short field landing over an obstacle (such as trees, or power lines), or to avoid an obstacle (such as a single tree on the extended centerline of the runway), and may be practiced as part of emergency landing procedures. These methods are also commonly employed when flying into farmstead or rough country airstrips where the landing strip is short. Pilots need to touch down with ample runway remaining to slow down and stop.

There are common situations where a pilot may deliberately enter a slip by using opposite rudder and aileron inputs, most commonly in a landing approach at low power.

Without flaps or spoilers, it is difficult to increase the steepness of the glide without adding significant speed. This excess speed can cause the aircraft to fly in ground effect for an extended period, perhaps running out of runway. In a forward slip much more drag is created, allowing the pilot to dissipate altitude without increasing airspeed, increasing the angle of descent (glide slope). Forward slips are especially useful when operating pre−1950s training aircraft, aerobatic aircraft such as the Pitts Special or any aircraft with inoperative flaps or spoilers.

Often, if an airplane in a slip is made to stall, it displays very little of the yawing tendency that causes a skidding stall to develop into a spin. A stalling airplane in a slip may do little more than tend to roll into a wing−level attitude. In fact, in some airplanes stalling characteristics may even be improved.

2. Forward−slip vs. Sideslip

Aerodynamically, these are identical once established, but they are entered for different reasons and will create different ground tracks and headings relative to those prior to entry. Forward−slip is used to steepen an approach (reduce altitude) without gaining much airspeed, benefiting from the increased drag. The sideslip moves the aircraft sideways (often, only in relation to the wind) where executing a turn would be inadvisable, drag is considered a byproduct. Most pilots like to enter sideslip just before flaring or touching down during a crosswind landing.

3. Sideslip

The sideslip also uses aileron and opposite rudder. In this case, it is entered by lowering a wing and applying exactly enough opposite rudder so the airplane does not turn (maintaining the same heading), while maintaining safe airspeed with pitch or power. Compared to Forward−slip, less rudder is used: Just enough to stop the change in the heading.

In the sideslip condition, the airplane's longitudinal axis remains parallel to the original flight path, but the airplane no longer flies along that track. The horizontal component of lift is directed toward the low wing, drawing the airplane sideways. This is the still−air, headwind or tailwind scenario. In case of crosswind, the wing is lowered into the wind, so that the airplane flies the original track. This is the sideslip approach technique used by many pilots in crosswind conditions (sideslip without slipping). The other method of maintaining the desired track is the crab technique: the wings are kept level, but the nose is pointed (part way) into the crosswind, and resulting drift keeps the airplane on track.

A sideslip may be used exclusively to remain lined up with a runway centerline while on approach in a crosswind or be employed in the final moment of a crosswind landing. To commence sideslipping, the pilot rolls the airplane toward the wind to maintain runway centerline position while maintaining heading on the centerline with the rudder. Sideslip causes one main landing gear to touch down first, followed by the second main gear. This allows the wheels to be constantly aligned with the track, thus avoiding any side load at touchdown.

The sideslip method for crosswind landings is not suitable for long–winged and low–sitting aircraft such as gliders, where instead a crab angle (heading into the wind) is maintained until a moment before touchdown. Aircraft manufacturer Airbus recommends the sideslip approach is arailiable only in low crosswind conditions.

飞行操纵性
Flight Maneuverability

Contents

1) Maneuverability

2) Lateral (pitching) maneuverability

3) Longitudinal (rolling) and vertical (yawing) maneuverability

4) Devices on main control surfaces

学习内容

1) 操纵性

2) 横向（俯仰）操纵性

3) 纵向（滚转）和垂直（偏航）操纵性

4) 操纵面上的附设装置

任务 1 操纵性
Task 1 Maneuverability

 Contents

Maneuverability

Learning Outcomes

1) Master the concept and classification of maneuverability

2) Master the meaning of aircraft maneuverability

3) Cultivate professional qualities of rigor, carefulness, and ability to express, coordinate, and communicate effectively

 任务内容

操纵性

任务目标

1）掌握操纵性的概念和分类

2）掌握飞机操纵性的含义

3）培养严谨、细心的职业素养，以及有效表达、协调和沟通的能力

Learning Guide

Aircraft maneuverability refers to the characteristic of an aircraft responding to pilot control. The pilot mainly controls the elevator, rudder and aileron through the control panel and pedals (Fig. 8–1) to change the aircraft from one flight state to another, so as to complete takeoff, climb, cruise, descent, approach and landing, etc. Control is one of the important flight qualities of an aircraft and an important aspect of flight mechanics research. The input of control is the force exerted by the pilot on the joystick or pedals, as well as the displacement of the joystick and pedals. The output is the changes in aircraft motion parameters, such as angle of attack, sideslip angle, tilt angle, various angular velocities, flight speed, altitude, and overload.

Fig. 8-1 Pilot control panel and peadals
图 8-1 飞行员控制面板及踏板

课文

Maneuverability

操纵性

Maneuverability refers to the characteristics of an aircraft changing from one flight state to another under the control of the pilot (Fig. 8–2).

The aircraft that is too sensitive or too slow to the pilot's control will bring difficulties to the flight control of the aircraft.

Being too sensitive to the control will make it difficult for the pilot to control the aircraft accurately, and it will also cause stall or structural damage due to excessive response to the control.

Being too slow to respond to the control makes the pilot has to increase the amount of control of the aircraft, which is very difficult to operate.

The stability and maneuverability should be balanced to make the aircraft have sufficient stability and good maneuverability.

操纵性是指在飞行员的控制下，飞机从一种飞行状态过渡到另一种飞行状态的特性（图 8–2）。

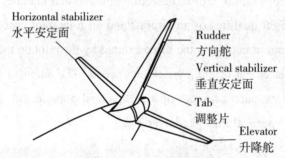

Fig. 8–2　Maneuverability and control surfaces of the aircraft

图 8–2　飞机的操纵性和操纵面

对飞行员的控制过于敏感或过于缓慢的飞机都给飞行控制带来了困难。

对控制过于敏感的飞机，不仅使飞行员难以准确地控制飞机，还会由于对控制的过度响应而导致失速或结构损坏。

对控制过于迟钝的飞机，使飞机无法及时对控制进行响应，飞行员必须增加飞机的控制量，这也给操作带来了困难。

稳定性和操纵性之间应保持平衡，以使飞机具有足够的稳定性和良好的操纵性。

1. The Lateral Maneuverability

1. 俯仰操纵性

The lateral maneuverability is the ability of the aircraft to rotate around the lateral axis, increase or decrease the angle of attack, and change the original flight attitude according to the pilot's control.

俯仰操纵性是指飞机根据飞行员的控制指令，围绕横轴转动增大或减小迎角，改变原始飞行姿态的能力。

2. The longitudinal maneuverability

2. 滚转操纵性

The longitudinal maneuverability is the ability of the aircraft to roll around the longitudinal axis and change the original flight attitude according to the pilot's control.

滚转操纵性是指飞机根据飞行员的控制指令，绕纵轴滚转，改变原始飞行姿态的能力。

3. The vertical maneuverability

3. 偏航操纵性

The vertical maneuverability is the ability of the aircraft to change its original flight attitude by rotating around the vertical axis and nosing left or right according to the pilot's control.

偏航操纵性是指飞机根据飞行员的控制指令，绕垂直轴偏航，向左或向右偏航来改变其原始飞行姿态的能力。

Maneuverability (1)   Maneuverability (2)

 New Words

maneuverability	[mə͵nuːvərə'bɪlɪti]	n.	操纵性
transit	['trænzɪt]	v.	经过；穿过
		n.	运输；运送；搬运；载运
accurately	['ækjʊrɪtli]	adv.	准确地；精确；正确

 Q&A

The following questions are for you to answer to assess the learning outcomes.

(1) Describe the definition of maneuverability.

(2) Describe the definition of lateral maneuverability.

(3) Describe the definition of vertical maneuverability.

(4) Describe the definition of longitudinal maneuverability.

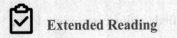

Control Systems for Large Aircraft

1. Mechanical Control

This is the basic type of system that was used to control early aircraft and is currently used in smaller aircraft where aerodynamic forces are not excessive. The controls are mechanical and manually operated.

The mechanical system of controlling an aircraft can include cables, push−pull tubes, and torque tubes. The cable system is the most widely used because deflections of the structure to which it is attached do not affect its operation. Some aircraft incorporate control systems that are a combination of all three. These systems incorporate cable assemblies, cable guides, linkage, adjustable stops, and control surface snubber or mechanical locking devices. These surface locking devices, usually referred to as a gust lock, limits the external wind forces from damaging the aircraft while it is parked or tied down.

2. Hydromechanical Control

As the size, complexity, and speed of aircraft increased, actuation of controls in flight became more difficult. It soon became apparent that the pilot needed assistance to overcome the aerodynamic forces to control aircraft movement. spring tabs, which were operated by the conventional control system, were moved so that the airflow over them actually moved the primary control surface. This was sufficient for the aircraft operating in the lowest of the high speed ranges (250–300 mph, 1mph=1.609, 344 km/h). For higher speeds, a power−assisted (hydraulic) control system was designed.

Conventional cable or push−pull tube systems link the flight deck controls with the hydraulic system. With the system activated, the pilot's movement of a control causes the mechanical link to open servo valves, thereby directing hydraulic fluid to actuators, which convert hydraulic pressure into control surface movements.

Because of the efficiency of the hydromechanical flight control system, the aerodynamic forces on the control surfaces cannot be felt by the pilot, and there is a risk of overstressing the structure of the aircraft. To overcome this problem, aircraft designers incorporated artificial feel systems into the design that provided increased resistance to the controls at higher speeds. Additionally, some aircraft with hydraulically powered control systems are fitted with a device called a stick shaker, which provides an artificial stall warning to the pilot.

3. Fly−by−wire Control

The Fly−by−wire (FBW) control system employs electrical signals that transmit the pilot's actions from the flight deck through a computer to the various flight control actuators. The FBW system evolved as a way to reduce the system mass of the hydromechanical system, reduce maintenance costs, and improve reliability. Electronic FBW control systems can respond to changing aerodynamic conditions by adjusting flight control movements so that the aircraft response is consistent for all flight conditions. Additionally, the computers can be programmed to prevent undesirable and dangerous characteristics, such as stalling and spinning.

Many of the new military high-performance aircraft are not aerodynamically stable. This characteristic is designed into the aircraft for increased maneuverability and responsive performance. Without the computers reacting to the instability, the pilot would lose control of the aircraft.

The Airbus A320 was the first commercial airliner to use FBW controls. Boeing used them in their Boeing 777 and newer design commercial aircraft. The Dassault Falcon 7X was the first business jet to use FBW control system.

任务 2　飞机的俯仰操纵性
Task 2　Lateral Maneuverability

Contents

　　1) Aircraft tail

　　2) Lateral control

　　3) Limit of center of gravity

Learning Outcomes

　　1) Master the method and principle of aircraft lateral control

　　2) Master the relationship between aircraft maneuverability and stability

　　3) Master the method and significance of determining the aircraft's CG

　　4) Be able to analyze the dynamics and flight quality problems in flight by using the lateral maneuverability of the aircraft

　　5) Cultivate professional qualities of rigor, carefulness, and ability to express, coordinate, and communicate effectively

任务内容

　　1）飞机的尾翼

　　2）俯仰操纵性

　　3）重心范围

 任务目标

　　1）掌握飞机俯仰操纵的方法和原理

　　2）掌握飞机操纵性和稳定性之间的关系

　　3）掌握飞机重心范围确定的方法和意义

　　4）能够运用飞机的俯仰操纵性分析飞行中的动力学和飞行品质问题

　　5）培养严谨、细心的职业素养，以及有效表达、协调和沟通的能力

Learning Guide

When we pull backward, the elevator will deflect upward for an angle, and the horizontal tail will generate downward additional lift, which will form a lifting moment for the center of gravity of the aircraft, which is called lateral moment. Under the action of torque, the aircraft begins to rotate around the horizontal axis, increasing the angle of attack. The increase in angle of attack causes the aircraft to generate additional upward lift, resulting in a stable lateral moment. Gradually, the two moments reach equilibrium, and the aircraft maintains stable flight at this angle of attack.

课文

1. Aircraft Tail
1. 飞机的尾翼

The horizontal tail (Fig. 8-3) of an aircraft is composed of a fixed (or adjustable installation angle) horizontal stabilizer at the front and an elevator that can deflect around the rotating shaft at the rear.

When the elevator deflects or adjust the trim angle of the horizontal stabilizer, the additional lift generated forms an additional lateral moment on the aircraft's CG, so as to maintain the lateral balance of the aircraft in different states and control the pitching of the aircraft.

飞机的水平尾翼（图 8-3）由位于前部的固定（或可调安装角度）的水平安定面和可绕后部转轴偏转的升降舵组成。

当升降舵偏转或调整水平安定面的配平角时，产生的附加升力在飞机重心上形成附加的俯仰力矩，从而在不同状态下保持飞机的俯仰平衡并控制飞机的俯仰。

264

Fig. 8-3　Horizontal tail

图 8-3　水平尾翼

2. Lateral Control

2. 飞机的俯仰操纵

The lateral control of the aircraft is accomplished by the pilot changing the deflection angle of the elevator through the control column or side stick in the flight compartment (Fig. 8-4).

The lateral moment generated on the horizontal stabilizer and elevators to the aircraft's CG is called the lateral control.

When the pilot pulls the control column backward, the elevators deflect upward, the additional lift generated on the horizontal tail is downward. The additional lateral moment generated forces the aircraft to raise up to increase its angle of attack, and reduce the flight speed.

When the pilot pushes the control column forward, the elevators deflect downward, the additional lift generated on the horizontal tail is upward. The additional lateral moment generated forces the aircraft to tilt down to reduce its angle of attack, and increase the flight speed.

飞机的俯仰操纵由飞行员通过操纵驾驶舱中的驾驶杆或侧杆改变升降舵的偏转角来完成（图 8-4）。

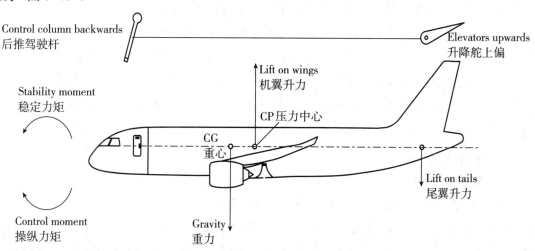

Fig. 8-4　Lateral control of the aircraft

图 8-4　飞机的俯仰操纵

由安定面和升降舵对飞机重心产生的俯仰力矩称为俯仰操纵力矩。

当飞行员向后拉杆时，升降舵向上偏转，水平尾翼上产生的附加升力向下。相对于重心产生的附加俯仰力矩使飞机抬头，飞机迎角增加，飞行速度降低。

当飞行员向前推杆时，升降舵向下偏转，水平尾翼产生的附加升力向上。相对于重心产生的附加俯仰力矩使飞机向下低头，飞机迎角减小，飞行速度增大。

3. Limit of the CG

3. 飞机的重心范围

The stability and maneuverability of the aircraft contradict each other.

If the aircraft is too stable, which means that it is hard to change the flight attitude and the aircraft is difficult to operate. If the aircraft is too sensitive, which means that it is easy to change the flight attitude and the aircraft is difficult to operate either.

The position of the CG of the aircraft has a great influence on the lateral static stability and maneuverability of the aircraft.

The position of the CG changes due to different loading, fuel consumption, and aircraft configuration.

To ensure that the aircraft has sufficient stability and maneuverability, the variation range of the aircraft's CG must be specified.

The changing limit of the CG is determined by the front limit of the CG and the rear limit of the CG. The change of the CG shall not exceed the range limited by the front and the rear limit of the CG.

飞机的稳定性和操纵性是相互制约的。

如果飞机稳定性太大，这意味着很难改变飞行姿态，飞机将很难操纵。如果飞机稳定性太小，这意味着很容易改变飞行姿态，飞机也同样很难操纵。

飞机重心的位置对飞机的俯仰稳定性和操纵性有很大影响。

由于不同的装载、油耗、飞机构型变化，重心位置也会发生变化。

为了确保飞机具有足够的稳定性和良好的操纵性，必须规定飞机重心的变化范围。

重心的变化范围由重心的前极限和后极限确定。重心的变化不得超过重心前极限和后极限所限制的范围。

1) Front Limit of the CG

1）重心前极限

The front limit of the CG is the position that allows the CG of the aircraft to be the most forward. As the CG of the aircraft moves forward, the lateral stability of the aircraft improves, but the lateral control moment required to change the original flight state of the aircraft is increased, so that the deflection angle and column force required to operate the aircraft are increased, which makes the aircraft too slow to respond to the control. The aircraft's maneuverability performance is reduced. In addition, if the CG of the aircraft is too forward , the required deflection angle of the elevtors may be too large to exceed the specified design value.

重心前极限是允许飞机重心最靠前的位置。随着飞机重心向前移动，飞机的俯仰稳定性增大，但改变飞机原始飞行状态所需的俯仰操纵力矩也相应增加，从而增加了操纵飞机所需的舵面偏转角和操纵行程，这使飞机对操纵的响应变得迟缓，飞机的操纵性能下降。此外，如果飞机重心太靠前，升降舵的所需偏转角度可能增大，超过设计的允许最大偏转值。

2) Rear Limit of the CG

2）重心后极限

The rear limit of the CG is the position that allows the CG of the aircraft to be farthest back. As the CG of the aircraft moves backward, the lateral stability of the aircraft decreases. The response of the aircraft to the control is too sensitive, and it is difficult to operate the aircraft accurately.

重心后极限是允许飞机重心最靠后的位置。随着飞机重心向后移动，飞机的俯仰稳定性降低，飞机对操纵的响应过于敏感，飞行员难以准确操作飞机。

| Lateral Maneuverability (1) | Lateral Maneuverability (2) | Lateral Maneuverability (3) | Lateral Maneuverability (4) |

 New Words

horizontal	[ˌhɒrɪ'zɒntl]	adj. 水平的；与地面平行的
stabilizer	['steɪbəlaɪzə(r)]	n. （飞机的）安定面；稳定装置
deflect	[dɪ'flekt]	v. 偏转；转移
elevator	['elɪveɪtə(r)]	n. （飞行器的）升降舵；电梯
tilt	[tɪlt]	v. 倾斜；（使）倾侧
rear	[rɪə(r)]	adj. 后方的；后面的

 Q&A

The following questions are for you to answer to assess the learning outcomes.

(1) Brief introduction of components of horizontal tail.

(2) How to maneuver an aircraft laterally?

(3) Briefly describe the relationship between lateral maneuverability and stability of aircraft.

(4) Why do we need to determine the front limit and rear limit of the center of gravity of an aircraft?

任务 3　飞机的滚转和偏航操纵
Task 3　Longitudinal and Vertical Maneuverability

 Contents

    1) Longitudinal and vertical control

    2) Adverse yaw

    3) Aileron reversion

    4) Spoilers

    5) Vortex generator

    6) Rudder

    7) Adverse roll

Learning Outcomes

    1) Master the principles and methods of aircraft longitudinal control

    2) Master the causes and elimination methods of harmful phenomena in aircraft longitudinal control

    3) Master the principles and methods of aircraft vertical control

    4) Master the causes and elimination methods of harmful phenomena in aircraft vertical control

    5) Be able to analyze the dynamics and flight quality problems in flight by using the longitudinal and vertical maneuverability of the aircraft

    6) Cultivate professional qualities of rigor, carefulness, and ability to express, coordinate, and communicate effectively

任务内容

    1）滚转和偏航操纵

    2）有害偏航

    3）副翼反逆

    4）扰流板

    5）涡流发生器

    6）方向舵

    7）蹬舵反倾斜

 任务目标

    1）掌握飞机滚转操纵的原理和方法

    2）掌握飞机滚转操纵中有害现象的产生原因和消除方法

    3）掌握飞机偏航操纵的原理和方法

    4）掌握飞机偏航操纵中有害现象的产生原因和消除方法

    5）能够运用飞机的滚转和偏航操纵性分析飞行中的动力学和飞行品质问题

    6）培养严谨、细心的职业素养，以及有效表达、协调和沟通的能力

Learning Guide

Same as flight stability, there is also mutual influence relationship between aircraft's roll and yaw operations. Usually we operate the ailerons to control the roll of the aircraft, and operate the rudder to control the yaw of the aircraft. In fact, controlling roll can also cause the aircraft to yaw, and controlling yaw can also cause the aircraft to roll. How is the specific process carried out? We will conduct in-depth exploration in this task.

课文

1. Longitudinal and Vertical Control

1. 飞机的滚转和偏航操纵

The longitudinal control of the aircraft is accomplished by deflecting the ailerons (Fig. 8-5). Ailerons are control surfaces mounted on the rotating shaft at the trailing edge of the wing.

When the pilot moves the side stick to the left (or turns the steering wheel to the left), the aileron on the left wing deflects upward and the lift of that wing decreases, the aileron on the right wing deflects downward and the lift of that wing increases. The moment generated by the asymmetric lift on the two wings makes the aircraft roll to the left (Fig. 8-6).

飞机的滚转控制是通过偏转副翼来完成的（图 8-5）。副翼是安装在机翼后缘旋转轴上的操纵面。

Neutral
中立位

Down
向下偏转

Up
向上偏转

Fig. 8-5　Ailerons
图 8-5　副翼

当飞行员向左移动侧操纵杆（或向左转动驾驶盘）时，左翼的副翼向上偏转，该翼的升力减小，右翼的副翼向下偏转，该机翼的升力增大。两翼不对称升力产生的力矩使飞机向左滚转（图 8-6）。

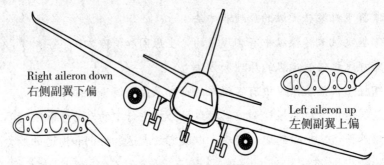

Right aileron down
右侧副翼下偏

Left aileron up
左侧副翼上偏

Fig. 8-6　Longitudinal control of the aircraft
图 8-6　飞机的滚转控制

2. Adverse Yaw

2. 有害偏航

Deflecting ailerons will not only produce longitudinal moment, but also vertical moment. Although the vertical moment is relatively small, it is adverse to the control of the aircraft, which is called adverse yaw.

The adverse yaw is mainly caused by the change of lift on two wings which are asymmetry, resulting in asymmetry drag on two wings.

When the side stick or control wheel are rolled left, the aileron on the left wing is deflected upward and the aileron on the right wing is deflected downward. With the lift change on both side of the wing, the drag on the right wing is larger than the drag on the left wing. The difference of the drags on the wings forces the aircraft to yaw right, which is adverse to the desired control.

When the aircraft yaws to the right around the vertical axis, there is left side slipping. Due to the longitudinal stability of the aircraft, the rolling moment generated by side slipping makes the aircraft roll to the right, which is opposite to the control purpose, reducing the control moment of rolling to the left, thereby reducing the control efficiency of ailerons.

Turning the side stick left to roll the aircraft left aims to make the aircraft to enter a circling to the left, but the vertical moment generated by the drag difference of the two wings makes the aircraft yaw right, which is also adverse to the horizontal turning control of the aircraft.

偏转副翼不仅会产生滚转力矩，还会产生偏航力矩。虽然偏航力矩相对较小，但对飞机的控制不利，称为有害偏航。

有害偏航主要是由两个机翼上的不对称升力变化，引起了两个机翼的不对称阻力造成的。

当操纵侧杆或驾驶盘操纵飞机向左滚转时，左翼副翼向上偏转，右翼副翼向下偏转。随着机翼两侧升力的变化，右翼上的阻力大于左翼上的阻力。机翼上的阻力差迫使飞机向右偏航，这对期望的左滚转操纵构成了有害偏航。

当飞机绕垂直轴向右偏航时，会发生左侧滑。由于飞机的纵向稳定性，侧滑产生的滚转力矩使飞机向右滚转，这与左滚转的操纵目的相反，减少了向左滚转时的操纵力矩，从而降低了副翼的操纵效率。

向左转动驾驶杆，使飞机向左滚转，目的是使飞机进入向左盘旋的状态，但两翼的阻力差产生的偏航力矩使飞机机头向右偏航，这也对飞机向左的水平转弯控制有害。

1) Differential Ailerons

1）差动副翼

In order to eliminate adverse yaw, differential ailerons can be used, whose ailerons are with upward deflection angle greater than downward deflection angle. This ailerons eliminate adverse yaw by generating more drag on the wing of the upper aileron to balance the drag of the wing on the other side.

为了消除有害偏航，可以使用差动副翼，其副翼向上偏转角度大于向下偏转角度。这种副翼通过在上偏副翼的机翼上产生更多阻力来平衡另一侧机翼的阻力，从而消除有害偏航。

2) Frise Ailerons

2）弗利兹副翼

Another ailerons that can be used to eliminate adverse yaw is the Frise ailerons (Fig. 8-7). The Frise ailerons move the rotation shaft of the ailerons backward from the leading edge of the ailerons and arrange the shaft on the lower surface of the aileron. When the aileron deflects downward, even if the maximum deflection angle is reached, the leading edge of the aileron will not expose through the upper surface of the wing. When the aileron deflects upward, even if the deflection angle is very small, the leading edge of the aileron will expose through the lower surface of the wing, resulting in large aerodynamic drag, so as to balance the drag on the lower side of the aileron and eliminate the adverse yaw caused by aileron deflection.

另一种可以用来消除有害偏航（图 8-7）的副翼是弗利兹副翼。弗利兹副翼将副翼的旋转轴从副翼的前缘向后移动，将轴安装在副翼的下表面上。当副翼向下偏转时，即使达到最大偏转角，副翼前缘也不会通过机翼上表面露出。当副翼向上偏转时，即使偏转角非常小，副翼前缘也会通过机翼下表面露出，从而产生较大的气动阻力，平衡副翼一侧偏侧的阻力，消除副翼偏转造成的有害偏航。

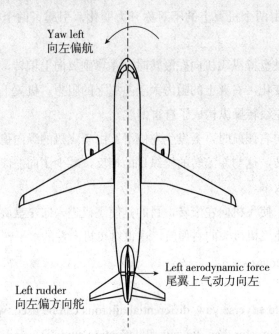

Fig. 8-7　Vertical maneuverability
图 8-7　飞机的偏航操纵控制

3) Aileron Reversion
3）副翼反逆

In flight, due to the elastic deformation of the wing, most of which is torsion deformation, the phenomenon that the ailerons lose their function or produce opposite effect, is called aileron failure or reversion. When the ailerons fail, this flight speed is called aileron reverse critical speed.

When the flight speed exceeds the aileron reverse critical speed, if the pilot pushes the side stick (or turns the steering wheel) to the left, the aircraft will roll to the right instead.

To improve the control efficiency of ailerons, prevent ailerons from reversing, and ensure flight safety, it is necessary to make the aircraft's flight speed less than the critical speed of ailerons.

Measures to improve the critical speed of ailerons are as below.

(1) Improve the torsional stiffness of the wing. The greater the torsional stiffness of the

wing, the smaller the torsional angle generated by the wing, and the higher the critical speed of ailerons.

(2) Another method is to use two groups of ailerons. Two groups of ailerons are arranged at the trailing edge of each wing, one group is near the wing tip, which is called the outer ailerons, the other group near the wing root is called inner ailerons. When the aircraft flies at low speeds, two groups of ailerons (or outboard ailerons) can be used to control the aircraft longitudinally, so as to improve the control efficiency of ailerons. When the aircraft flies at high speeds, only the inner ailerons are used for longitudinal control of the aircraft.

在飞行中，由于机翼弹性变形（多为扭转变形）的影响，副翼完全失去其功能或产生相反的效果，这称为副翼失效或反效。副翼失效时的飞行速度称为副翼反效临界速度。

当飞行速度超过副翼反逆临界速度时，若飞行员向左推动侧杆（或转动驾驶盘）时，飞机将向右滚转。

为了提高副翼的控制效率，防止副翼反效，确保飞行安全，有必要使飞机飞行速度低于副翼反效临界速度。

提高副翼临界速度的措施如下。

（1）提高机翼的扭转刚度。机翼的扭转刚度越大，机翼产生的扭转角越小，副翼反效的临界速度越高。

（2）另一种方法是使用两组副翼。两组副翼安装在每个机翼的后缘：一组靠近翼尖称为外副翼，另一组靠近翼根称为内副翼。当飞机低速飞行时，可使用两组副翼（或外侧副翼）操纵飞机的滚转，以提高副翼的操纵效率。当飞机高速飞行时，只有内副翼用于飞机的滚转操纵。

4. Spoilers

4. 扰流板

The spoilers are rectangular plates whose front edge is a hinge shaft that arranged on the upper surface of the wing and the front edge of the flap.

They are attached to the wing to form a smooth surface fairing shape, when the spoilers are not working. They rotate around the rotating shaft and opens upward and forward, forming an angle with the wing upper surface during its operation.

When the aircraft is operated longitudinally, if the aileron deflection angle reaches a certain value, the spoilers at the same side with the upper aileron deflect upward by the linkage mechanism. When the spoilers deployed, the airflow separation behind the spoilers further reduce the lift of that wing to increase the longitudinal moment, and improve the longitudinal control efficiency of the ailerons.

The ground spoilers are near the fuselage and the flight spoilers are on the outside. When the aircraft flies, the ground spoilers are locked, the flight spoilers assist the aileron to complete the longitudinal control of the aircraft, or reduce the flight speed by extending flight spoilers on both side.

When landing, as soon as the wheel touches the ground, the ground spoilers are unlocked, and all spoilers on the wings are opened to reduce lift and increase drag to shorten the landing and running distance of the aircraft on the runway.

扰流板为矩形板，其前缘为铰接轴，安装在机翼上表面和襟翼前缘。

当扰流板不工作时，它们向机翼回收，和机翼一起形成光滑的流线型。工作时，扰流板绕转轴旋转，向上和向前打开，与机翼上表面形成一定角度。

当操纵飞机进行滚转时，当副翼偏转角达到一定值时，与向上偏转的副翼同一侧的扰流板通过连接机构向上偏转。当扰流板展开时，扰流板后面分离的气流进一步降低了该侧机翼的升力，提高了副翼的滚转控制效率。

地面扰流板靠近机身，飞行扰流板位于机身外侧。当飞机飞行时，地面扰流板被锁定，飞行扰流板帮助副翼完成飞机的滚转控制，或者通过同时打开两侧的飞行扰流板来降低飞行速度。

着陆时，在机轮接地瞬间，地面扰流板解锁，机翼上的所有扰流板打开，以减少升力和增加阻力，缩短飞机在跑道上的着陆和滑跑距离。

5. Vortex Generator

5. 涡流发生器

As stated in moudles 5, vortex generator is a device that uses vortex to bring energy into the boundary layer from the external airflow to accelerate the airflow in the boundary layer, and prevent air separation.

Some aircraft often arrange vortex generators on the upper surface of the wing and in front of the ailerons, which can effectively delay the boundary layer separation and improve the aileron control efficiency at large deflection angle and high speed.

正如模块 5 中所述，涡流发生器是一种利用涡流将外部气流的能量带入附面层，加速附面层内气流流动并防止空气分离的装置。

通常，在飞机机翼上表面和副翼前方布置涡流发生器，可有效延缓附面层的气流分离，并在大偏转角和高速的情况下提高副翼的操纵效率。

6. Rudder

6. 方向舵

The rudder is the control surface mounted on the fin. The fin is composed of a vertical stabilizer and a rudder. The stabilizer is fixed on the fuselage, and the rudder is hinged on the rotating shaft at the rear edge of the stabilizer.

The pilot can control the rudder to deflect left and right through pedals to control the heading of the aircraft.

方向舵是安装在垂尾上的操纵面。垂尾由垂直安定面和方向舵组成。安定面固定在机身上，方向舵铰接在安定面后缘的转轴上。

飞行员可以通过脚蹬控制方向舵左右偏转，以控制飞机的航向。

7) Adverse Roll

7）蹬舵反倾斜

When the rudder deflects, one part of the longitudinal moment is opposite to the desired direction of aircraft rolling, this phenomenon is the adverse roll.

When pedaling the left rudder, the longitudinal moment produced by the lateral force on the fin makes the aircraft head to the right.

In this way, the left hand yaw will make the aircraft head to the right, and the right hand yaw will make the aircraft head to the left.

当方向舵偏转时，飞机上一部分的滚转力矩与所需的飞机滚转方向相反，这种现象称为蹬舵反倾斜。

当踩下左方向舵脚蹬时，机翼上的侧向力产生的滚转力矩使飞机向右滚转。

这样，左偏航将使飞机向右滚转，右偏航将让飞机向左滚转。

Longitudinal and Vertical Maneuverability (1)

Longitudinal and Vertical Maneuverability (2)

Longitudinal and Vertical Maneuverability (3)

Longitudinal and Vertical Maneuverability (4)

Longitudinal and Vertical Maneuverability (5)

Longitudinal and Vertical Maneuverability (6)

Longitudinal and Vertical Maneuverability (7)

Longitudinal and Vertical Maneuverability (8)

Longitudinal and Vertical Maneuverability (9)

Longitudinal and Vertical Maneuverability (10)

Longitudinal and Vertical Maneuverability (11)

New Words

pilot	['paɪlət]	n.	飞行员
asymmetric	[ˌeɪsɪ'metrɪk]	adj.	不对称；不对等的
adverse	['ædvɜːs]	adj.	不利的；有害的；反面的
thereby	[ˌðeə'baɪ]	adv.	因此；由此；从而
differential	[ˌdɪfə'renʃl]	adj.	有差别的；差别的

 Q&A

The following questions are for you to answer to assess the learning outcomes.

(1) How does an airplane maneuver longitudinally?

(2) Describe the definition of adverse yaw.

(3) How to eliminate adverse yaw?

(4) How does an airplane maneuver vertically?

(5) Describe the definition of aileron failure or reversion.

(6) How to increase the critical velocity of aileron?

(7) How to improve the efficiency of longitudinal control?

(8) How does the rudder maneuver vertical control?

 Extended Reading

Adverse Yaw—What Is It?

We have experienced a good amount of non-flyable weather for the past 10 days as of writing this article, providing a lot of ground school time with students. One weak subject area encountered is knowing and understanding adverse yaw. It is not often well understood by general aviation pilots. What is it? How does it occur? What can be done to correct it? What happens if nothing is done to correct it?

First, let's define yaw. According to the Merriam-Webster online dictionary, yaw can be explained as a side-to-side movement or, specific to aviation, to turn by angular motion about the vertical axis. Adverse yaw is the tendency for the nose of an airplane to yaw in the opposite direction when an airplane banks its wings for a turn. The increased lift of the raised wing results in the increased drag, which causes the airplane to yaw or swing toward the side or direction of the raised wing. The rudder is typically used to counteract adverse yaw.

If uncorrected while in normal flight, safety is not usually affected—but it can lead to a more serious situation. Rather, it generally creates inefficiency in flight and possibly physical discomfort among passengers as the aircraft skids and slips from side to side while conducting uncoordinated turns.

1. When does Adverse Yaw Occur?

Adverse yaw can occur in many situations but is most often experienced when making a turn while in flight. For instance, when a pilot initiates a turn to the left, the yoke or control column is rotated or moved leftward. That causes the left aileron to rise upward, pushing the left wing down. While this is occurring, simultaneously the right aileron travels downward, generating more lift and forcing the right wing upward. Generating more lift also generates more drag, so in this left turn, the right wing pulls or "drags" the upward wing away from the direction of the turn. That causes the nose to move or yaw to the right before turning left. This is adverse yaw.

The yaw is most evident in slower airplanes with long wings, like J–3 Cubs, Taylorcrafts, Aeroncas, and Cessna trainers. Many aircraft manufacturers later designed the ailerons to help offset the yaw, but the primary control for efficiently managing the yaw continues to be the rudder.

Some students, as well as pilots, have a difficult time recognizing adverse yaw. When this occurs, I demonstrate it by establishing straight and level flight at a speed somewhat slower than the normal cruising speed. The slower the aircraft is flying, the more pronounced the yaw. Rotate the control yoke to the left or move the control column to the left, establishing a banking left turn. Do not make any rudder inputs. As the right wing rises, the nose will move or yaw to the right in a rather pronounced movement before finally moving to the left. Return to level flight using ailerons only, and the yaw is again clearly visible. Now, try a turn to the right without adequate rudder and watch the nose move to the left or away from the turn. The adverse yaw is quite pronounced.

2. Where does Adverse Yaw Occur?

Most flight training manuals explain adverse yaw by describing banking turns as mentioned above. However, it can occur in several other flight attitudes, some quite critical.

A crosswind takeoff creates adverse yaw for instance, but it is not usually recognized by most pilots. When setting up for a crosswind takeoff with a wind from left to right, the control column is positioned to a full left deflection. The left aileron is deflected full up to prevent the left wing from flying before the right wing. The right aileron is full down, creating increased lift —but also more drag. As groundspeed increases during the takeoff roll, the right wing is actually helping the aircraft remain aligned with the centerline by creating both lift and drag. But the aileron does not create enough drag to offset the crosswind, nor the engine torque and propeller P–factor being generated to keep the airplane moving straight down the runway. In this example, right rudder must also be applied to help maintain directional control.

In a common crosswind, as ground speed increases the ailerons become increasingly more effective. Less aileron correction is needed to keep one wing from flying before the other. The lesser the down aileron, the lesser adverse yaw is being created. Just at the instant of takeoff, the ailerons should be moved to the neutral position to eliminate the adverse yaw.

Climbing turns create adverse yaw, especially when making right turns. But is it critical? Usually not, but it does create inefficiency and a lack of good coordinated control inputs.

3. Adverse Yaw in the Traffic Pattern

As a flight instructor, one area in which I observe adverse yaw situations is in the traffic pattern in preparation for landing. When abeam the runway numbers on the downwind leg, power is reduced to begin slowing the aircraft. Remember practicing slow flight? The slower the airspeed, the more sluggish the controls, which then require more input.

Initiating the descending left turn onto the base leg at this slower airspeed requires a fair amount of left column movement. Frequently, the pilot does not match the increased aileron input with coordinated left rudder input. For the first few seconds, the airplane is in a significant adverse yaw situation with the nose moving to the right while the airplane is trying to turn left. Slowly the nose begins to follow the rest of the plane and sluggishly turns left. About halfway through the turn, the aircraft stabilizes, completing the turn. At the point of rollout leveling the wings on the base leg, increased right stick needs to be applied. Again, the extra right rudder input is seldom applied. The aircraft is again placed in an adverse yaw situation as the airplane turns right but the nose moves leftward. This situation has become a bit more critical because the airspeed has properly been reduced.

Finally, at yet a slower speed, the aircraft banks left onto the final approach leg aligning with the runway in preparation for the landing. In most light aircraft, the airspeed has been reduced another 10 m/h, making aileron and rudder inputs even more sluggish. When the left turn onto final is initiated, significant left stick movement and left rudder application are again needed for the coordinated turn. However, that is often not what I observe. Rather, the proper aileron inputs are made but not the rudder. Again, the aircraft is now in an adverse yaw configuration and at a slow airspeed. Safety is now compromised, especially if the pilot allows the nose to pitch upward. The aircraft is in a near cross–control configuration. This is the critical area where stalls and stall/spins can occur.

Coordinated aileron and rudder inputs will eliminate a good portion of this critical situation, preventing adverse yaw. However, a second critical situation is often created if the pilot overshoots the runway on the turn to final. Rather than realigning using coordinated aileron and rudder, the controls are cross-controlled. Hard left rudder is applied in an attempt to bring the aircraft back in the line with the runway centerline, but right aileron is also being applied so as not to create a steep bank. This configuration can be quite safe if attempting to perform a slip and the pilot knows how to execute it properly. However, this uncoordinated situation is a step toward disaster. Cross-controlling unintentionally causes the nose to pitch upward and the airspeed to dissipate. The aircraft is now in a more critical stall or stall/spin configuration while close to the ground.

An astute pilot will be aware of the control input needs and fly the airplane in a coordinated manner. A lackadaisical pilot will be oblivious to the added inputs needed when the aircraft is flown at a slower airspeed, creating the adverse yaw situations I've described. This compromises safety for both the pilot and the passengers enjoying a flight in a small general aviation aircraft.

Be aware, practice good coordination, and fly safely so that you can enjoy a flight again on another day.

任务 4　飞机操纵面上的附设装置
Task 4　Devices on Main Control Surfaces

Contents

1) Devices attached on main control surfaces

2) Mass balance

3) Aerodynamic compensation

4) Aerodynamic balance

5) Tab

6) Horizontal stabilizer

Learning Outcomes

1) Master the types and uses of the devices attached on the aircraft control surfaces

2) Master the working principle of attached devices on the aircraft control surfaces

3) Be able to analyze the dynamics and flight quality problems in flight by using the attached devices on the aircraft control surfaces

4) Cultivate professional qualities of rigor, carefulness, and ability to express, coordinate, and communicate effectively

 任务内容

  1）飞机操纵面上的附设装置

  2）质量平衡

  3）气动补偿

  4）气动平衡

  5）调整片

  6）水平安定面

 任务目标

  1）掌握飞机操纵面上附设装置的种类和用途

  2）掌握飞机操纵面上附设装置的工作原理

  3）能够运用飞机操纵面上的附设装置分析飞行中的动力学和飞行品质问题

  4）培养严谨、细心的职业素养，以及有效表达、协调和沟通的能力

Learning Guide

  Flight controls maneuver the aircraft by manipulating different flight control surfaces. As we know before, flight control surfaces include ailerons elevators and rudder. With the continuous improvement of aircraft flight speed, greater driving force is required for the control of control surfaces, and the control efficiency of the surfaces also affects the flight efficiency. Therefore, people use many methods to improve the operational efficiency of flight control surfaces, such as, balancing the mass of control surfaces, and preventing phenomena such as chopping and shaking.

课文

1. Devices on Main Control Surfaces

1. 飞机操纵面上的附设装置

The functions of the attached devices on the main control surfaces of the aircraft are mass balance, aerodynamic compensation and aerodynamic balance.

飞机主要操纵面上的附设装置的功能有质量平衡、空气动力补偿和空气动力平衡。

2. Mass Balance
2. 质量平衡

Mass balance (Fig. 8-8) is to add a counterweight inside the leading edge of the control surface to move the center of gravity of the control surface forward and in front of the rotating shaft. The purpose is to prevent the flutter of the control surface.

质量平衡（图 8-8）方法是在操纵面前缘内添加一个配重，从而将操纵面重心向前移动到旋转轴前方。其目的是防止操纵面颤振。

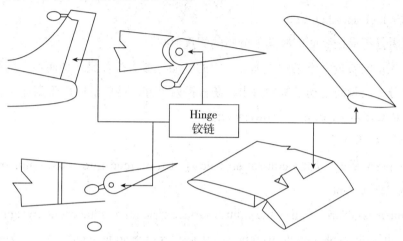

Fig. 8-8　Typical mass balance method
图 8-8　典型的质量平衡方法

3. Centralized Counterweight
3. 集中配重

The centralized counterweight uses counterweights that are installed far away from the rotating shaft at the leading edge of the control surface through strut or other supporting fittings.

This centralized counterweight uses less mass to makes the center of gravity of the control surface move forward significantly. However, since the counterweight often protrudes in the airflow, it will increase aerodynamic drag, so it does not meet the requirements of dynamic balance in the continuous change of the mass of the control surface along the spanwise. So its anti flutter effect is poor.

集中配重指通过支架或其他支撑配件安装在距传动轴较远的操纵面前缘处的可使用配重。

这种集中配重可以使用较少的质量使操纵面的重心显著向前移动。由于在气流中突出出来的配重会增加气动阻力，并且操纵面质量沿翼展方向的变化不满足动平衡要求，因此其抗颤振效果较差。

4. Distributed Counterweight

4. 分散配重

Distributed counterweight is to distribute the counterweight along the front edge of the control surface.

The mass of distributed counterweight is more than centralized counterweight. But its aerodynamic shape is well faired and will not increase the drag. Moreover, the mass of each section of the control surface is balanced, and the flutter prevention effect is good, which is widely used in high speed aircraft.

分布配重是指沿操纵面前缘均匀分布的配重形式。

分布式配重的质量大于集中式配重。它的气动外形更符合流线型而不会增加阻力。且操纵面各部分的质量满足动平衡的要求，防颤振效果好，被广泛应用于高速飞机。

5. Fixed and Adjustable Counterweight

5. 固定和可调配重

The counterweight of the control surface is also divided into fixed counterweight and adjustable counterweight.

When the mass distribution of the control surface changes and the counterweight needs to be re-balanced, the adjustable counterweight can be used for the adjustment.

The mass and location of the fixed counterweight are unchanged.

操纵面配重分为固定配重和可调配重。

当操纵面质量分布发生变化且需要重新平衡配重时，可使用可调配重进行调整。

固定配重的质量和位置不变。

6. Hinge Moment

6. 铰链力矩

The moment of aerodynamic force acting on the control surface to the rotating shaft of it is called the hinge moment.

When deflecting the control surface, the moment that overcomes the hinge moment to deflect the control surface is called the control moment.

铰链力矩指作用在操纵面上的气动力对其转轴的力矩。

当偏转操纵面时，克服铰链力矩使操纵面偏转的力矩称为操纵力矩。

7. Aerodynamic Compensation

7. 气动补偿

Aerodynamic compensation (Fig. 8-9) refers to the method of adding aerodynamic compensation devices to the control surface in accordance with the principles of aerodynamics, in order to reduce the hinge moment of the control surface and to easy the pilots' operation.

With the increase of aircraft speed nowadays, the hinge moment increases rapidly, and the pilot's manpower to control the device in the flight compartment also increases.

The purpose of aerodynamic compensation is to reduce the hinge moment and reduce the manpower of pilots in controlling the aircraft.

气动补偿（图8-9）是指为了减小操纵面的铰链力矩，使驾驶员操纵飞机时省力，按照空气动力学的原理对操纵面增加气动补偿装置的方法。

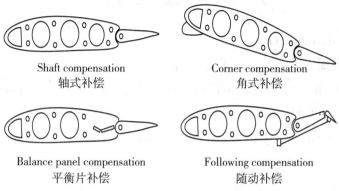

Shaft compensation
轴式补偿

Corner compensation
角式补偿

Balance panel compensation
平衡片补偿

Following compensation
随动补偿

Fig. 8-9 Typical aerodynamic compensation methods
图8-9 典型的气动补偿装置和方法

随着当代飞机速度的增加，铰链力矩也增加，飞行员控制驾驶舱内的控制装置的人力也随之增加。

气动补偿的目的是减少铰链力矩，降低飞行员操纵飞机的工作强度。

8. Aerodynamic Balance

8. 气动平衡

When the aircraft is in a certain flight state, aerodynamic balance aims to completely eliminate the force on the control device, so as to achieve un-holding column flight.

There are two kinds of aerodynamic balancing devices, tab and horizontal stabilizer with variable installation angle.

气动平衡是指当飞机处于某一飞行状态时，完全消除驾驶杆力，从而实现松杆飞行。

有两种气动平衡装置：调整片和安装角度可变的水平安定面。

1) Tab

1）调整片

The triming tab is a small wing surface installed behind the control surface, which can rotate around the hinge shaft. The pilot uses the trim handle or switch to drive the motor or screw to control the tab (Fig. 8-10).

调整片是安装在舵面后缘的一个小翼面，可绕铰接轴转动。飞行员使用配平手轮或电门以驱动电机或螺杆来控制调整片（图8-10）。

Elevator up with tab down
For nose up trimming
升降舵上偏,
调整片下偏进行抬头配平

Elevator down with tab up
For nose down trimming
升降舵下偏,
调整片上偏进行低头配平

Fig. 8-10 Trimming tab
图 8-10 可配平的调整片

2) Horizontal Stabilizer

2）水平安定面

Horizontal stabilizer with variable installation angle is a kind of aerodynamic balancing device commonly used in high speed large aircraft at present (Fig. 8-11).

Its rear beam is fixed on a hinge support, and the lower part of the front beam is connected with an actuator through universal joints.

The pilot can raise or lower the joint of the front beam through the balancing hand wheel and then the actuator makes the stabilizer to rotate around the hinge joint of the rear beam, hence to change its installation angle for aerodynamic balance purpose.

安装角度可调的水平安定面是目前高速大型飞机上常用的一种气动平衡装置（图 8-11）。

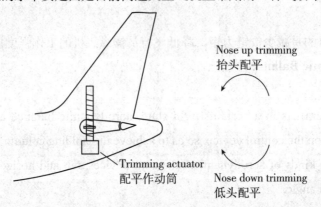

Nose up trimming
抬头配平

Trimming actuator
配平作动筒

Nose down trimming
低头配平

Fig. 8-11 Horizontal stabilizer with variable installation angle
图 8-11 安装角可调的水平安定面

水平安定面的后梁固定在铰接支架上，前梁下部通过连接结构与作动筒结构连接。

飞行员通过配平手轮使前梁的接头升高或降低，使安定面绕后横梁的铰接接头转动，改变其安装角度，实现气动平衡。

9. Aerodynamic Balance Vs. Compensation

9. 气动平衡与补偿的比较

The purposes of the aerodynamic balance and the aerodynamic compensation are different.

Aerodynamic balance is to completely counteract the hinge moment after the aircraft reaches a certain flight state, then the pilot releases the control lever, and the aircraft maintains its flight state.

Aerodynamic compensation is to reduce the hinge moment and reduce the manpower of the pilot when the pilot deflects the control surface and controls the aircraft.

The operation mode of the aerodynamic balance and the aerodynamic compensation are different.

The aerodynamic balance device does not work with the deflection of the control surface, but is operated by the pilot through an independent trim hand wheel or trim switch.

气动平衡和气动补偿的目的不同。

气动平衡是在飞机达到某一飞行状态后完全抵消铰链力矩，然后飞行员释放操纵杆，飞机仍保持飞行状态。

气动补偿是为了在飞行员偏转操纵面并操纵飞机时，减少铰链力矩，降低飞行员的工作强度。

气动平衡和气动补偿的操作模式不同。

气动平衡装置不随操纵面偏转而工作，而是由飞行员通过独立的配平手轮或配平开关操作。

Device on Main Control Surfaces (1)

Device on Main Control Surfaces (2)

Device on Main Control Surface (3)

Device on Main Control Surfaces (4)

Device on Main Control Surfaces (5)

Device on Main Control Surfaces (6)

 New Words

| mass | [mæs] | n. | 质量 |
| counterweight | ['kaʊntəweɪt] | n. | 配重；平衡物 |

flutter	['flʌtə(r)]	v.	颤振；（使）飘动；挥动
centralize	['sentrəlaɪz]	v.	集权控制；实行集中
spanwise	[spænwaiz]	adj.	顺翼展方向的
compensation	[,kɒmpen'seɪʃn]	n.	补偿；赔偿；补偿
trim tab	[trɪm tæb]		调整片

 Q&A

The following questions are for you to answer to assess the learning outcomes.

(1) What is the function of the devices attached to the aicraft control surfaces?

(2) Describe the definition of mass balance.

(3) Describe the definition of centralized counterweight.

(4) Describe the definition of distributed counterweight.

(5) What is the purpose of aerodynamic compensation?

(6) What are the aerodynamic compensation devices?

(7) Describe the definition of aerodynamic balance.

(8) What are the aerodynamic balance devices?

 Extended Reading

Horizontal vs. Vertical Stabilizers in Airplanes: What's the Difference?

Stabilizers are an important components of an airplane. They live up to their namesake by "stabilizing" the airplane and, thus, preventing unwanted movement. Whether it's a commercial jet or a private propeller airplane, most airplanes are designed with stabilizers. There are two primary types of stabilizers used in airplanes, however, including horizontal and vertical. So, what's the difference between horizontal and vertical stabilizers exactly?

1. What Is a Horizontal Stabilizer?

Located on the left and right sides of the airplane's tail, a horizontal stabilizer is designed to maintain the airplane's trim, in works by creating an upwards force that balances the airplane, horizontally, during flight. As the airplane flies, its horizontal stabilizers will push the air upwards to prevent swings in trim.

Horizontal stabilizers are rather simple components that consist of small and thin pieces of material—typically the same material from which the fuselage is constructed. They essentially look like small wings on the sides of the tail. Both the left and right sides of an airplane's tail will have a horizontal stabilizer. While horizontal stabilizers create a vertical force during flight, they extend horizontally from the sides of the tail.

2. What Is a Vertical Stabilizer?

Also known as a vertical fin, a vertical stabilizer is a component that's designed to minimize side slip—a phenomenon that occurs when an airplane begins to fly to the side—while subsequently helping the airplane maintain its course. Without a vertical stabilizer, an airplane may be pushed to the side. When side slip such as this occurs, it can take the airplane off its intended course. Pilots can typically correct their course, but it comes at the cost of additional fuel consumption. Therefore, vertical stabilizers are used to solve these problems.

The vertical stabilizer is found on the tail of an airplane. Like horizontal stabilizers, it consists of a small and thin piece of material that looks like a miniature wing. The vertical stabilizer, however, is installed vertically on the airplane's tail, whereas the horizontal stabilizers are installed horizontally on the airplane's tail.

It's important to note that some airplanes have multiple vertical stabilizers. Several military airplanes, for instance, feature either two or three vertical stabilizers. Known as twin-tail or triple-tail wings, respectively, they are designed to offer a superior level of control. With that said, most commercial airplanes only have a single vertical stabilizer. When looking at a commercial airplane, you'll see the vertical stabilizer located on the top of the tail.

参 考 文 献
References

［1］何庆芝．航空航天概论［M］．北京：北京航空航天大学出版社，1997.

［2］宋静波．飞机构造基础［M］．北京：航空工业出版社，2004.

［3］刘沛清．空气动力学［M］．北京：科学出版社，2023.

［4］谢础，贾玉红．航空航天技术概论［M］．北京：北京航空航天大学出版社，2005.

［5］刘得一，张兆宁，杨新湦．民航概论［M］．北京：中国民航出版社，2000.